MAPLE V
PRIMER
Release 4

Frank Garvan

CRC Press
Taylor & Francis Group
Boca Raton London New York

CRC Press is an imprint of the
Taylor & Francis Group, an **informa** business

CRC Press
Taylor & Francis Group
6000 Broken Sound Parkway NW, Suite 300
Boca Raton, FL 33487-2742

ISBN 13: 978-0-8493-2681-3 (pbk)
ISBN 13: 978-1-138-44247-4 (hbk)

Visit the Taylor & Francis Web site at
http://www.taylorandfrancis.com

and the CRC Press Web site at
http://www.crcpress.com

Library of Congress Cataloging-in-Publication Data

Garvan, Frank (Frank G.), 1955-
 The Maple V primer : release 4 / Frank Garvan.
  p. cm.
 Includes bibliographical references and index.
 ISBN 0-8493-2681-8 (alk. paper)
 1. Maple (Computer file). 2. Algebra—Data processing. I. Title.
QA155.7.E4G37 1996
510′.285′53—dc20   96-34579
                 CIP

Library of Congress Card Number 96-34579

About the author

Frank G. Garvan was born in Sydney, Australia in 1955. He obtained an Honours B.Sc. in mathematics in 1977 from the University of New South Wales. After a stint as a high-school teacher in outback N.S.W., he returned to the University of New South Wales to do an M.Sc. under the supervision of Michael Hirschhorn. In 1986, he obtained a Ph.D. at the Pennsylvania State University under the supervision of George Andrews. He has had post-doctoral positions at the University of Wisconsin, University of Minnesota, Macquarie University, and Dalhousie University. While at the University of Minnesota in 1988, he was introduced to Maple and has been a devotee ever since. He has been on the faculty at the University of Florida since 1990 and is now Associate Professor of Mathematics. His research interests include number theory, q-series, theory of partitions, and special functions. He enjoys incorporating Maple in both his teaching and research.

INTRODUCTION

MAPLE V® is a very powerful interactive computer algebra system. It is used by students, educators, mathematicians, scientists, and engineers for doing numerical and symbolic computation. MAPLE V has many strengths: (1) it can do exact integer computation, (2) it can do numerical computation to any (well almost) number of specified digits, (3) it can do symbolic computation, (4) it comes with many built-in functions and packages for doing a wide variety of mathematical tasks, (5) it has facilities for doing 2- and 3-dimensional plotting and animation, (6) it has a worksheet-based interface, (7) it has facilities for making technical documents, and (8) MAPLE V is a simple programming language which means the user can easily write his/her own functions and packages.

The purpose of this book is to help you get started using the main features of MAPLE V (Release 4), the latest version of Maple. It is not an exhaustive manual. The reader should consult the MAPLE V reference books when the need arises. It is best to use this book while at the computer trying out commands, following examples, and ex-

---

® Maple is a registered trademark of Waterloo Maple Inc., 450 Phillip Street, Ontario, Canada N2L 5J2, 1-800-267-6583, (519) 747-2373, Fax: (519) 747-5284, Email: info_web@maplesoft.com,
Website: http://www.maplesoft.com.

perimenting as you read. This book should be a sufficient resource for students taking a class that uses MAPLE V.

MAPLE V itself has built-in help facilities. Help can be found either through the interactive Help menu or by using the ? command. For instance, a very short introduction to MAPLE V can be found by typing ?intro.

The table of contents is organized mainly according to mathematical topics starting with using MAPLE V as a calculator, then doing high school algebra, calculus, and progressing to more sophisticated mathematics and programming. An important goal of this book is to show you how to write simple MAPLE V programs (or procedures). When this goal is achieved, the reader should appreciate the power of MAPLE V.

In Chapter 1 there is a brief introduction to the new MAPLE V interface. The reader anxious to know more about MAPLE V's new document creating facilities can find detailed information in Chapter 10.

MAPLE V is available on Windows, MacIntosh, and UNIX platforms. The author would like to thank Waterloo Maple Inc. for permission to include pictures of the maple icons and buttons.

Frank Garvan (frank@math.ufl.edu)
Department of Mathematics, University of Florida, Gainesville, FL 32611-8105

Contents

1. GETTING STARTED

1.1 Starting a MAPLE V session

In most systems a MAPLE V session is started by double clicking on the MAPLE V logo

Double clicking on the **New User's** Tour icon gives you a brief introduction to MAPLE V and its worksheet environment. In the X Windows version, MAPLE V is started by entering the command **xmaple**. In the text (tty) version, the Maple logo appears followed by the > prompt.

In most versions a window with menus will appear. At the top are the menus File, Edit, View, Format, Options, and Help. Beneath are two rows of buttons. The first row of buttons is called the *tool bar* and contains 19 buttons:

Create a new worksheet.

Open an existing worksheet.

Save the active worksheet.

Print the active worksheet.

Cut the selection to the clipboard.

Copy the selection to the clipboard.

Paste the clipboard contents into the current worksheet.

1

⟲	Undo the last deletion.
Σ	Insert maple commands.
T	Insert text.
[>	Insert a new maple input region after the cursor.
⇆	Convert the selected subsection into a section.
⇥	Convert the selection into a subsection.
⏹	Interrupt the current computation.
x	Set magnifcation to 100%.
𝑥	Set magnifcation to 150%.
𝑿	Set magnifcation to 200%.
¶	Display non-printing characters.
✛	Resize the active window to fill the available space.

The next row is called the *context bar* and contains four buttons:

x	Toggle the expression display between mathematical and maple notation.
🍁	Toggle the expression display between inert text and executable maple command.
(/)	Auto-correct the expression for syntax.

2

Execute the current expression.

The > prompt will be in the *worksheet* window.
Don't worry about the buttons too much at this
stage.

1.2 Basic syntax

In MAPLE V the prompt is the symbol >.
MAPLE V commands are entered to the right of
the prompt. Each command ends with either ":"
or ";". If the colon is used, the command is ex-
ecuted but the output is not printed. When the
semi-colon is used, the output is printed. Try typing 105/25: followed by a return (or enter).

```
>   105/25:
```

Observe that the output was not printed. Now
type 105/25;

```
>   105/25;
```
$$21/5$$

Try these

```
>   105/25-1/5;
```
$$4$$

```
>   "+1/5;
```
$$21/5$$

```
>   "";
```
$$4$$

3

Observe that MAPLE V uses exact arithmetic. The double quote character " refers to the previous result. The two double quotes "" refers to the result before the previous result. It is possible to refer back 3 lines using """, but one cannot refer back any further.

One of the most common mistakes is to omit the semi-colon or colon:

```
>   105/25
Warning, incomplete statement or missing
semicolon
>   105/25;
syntax error, unexpected number
```

Don't panic! MAPLE V has interpreted this to mean 105/25 105/25 hence the syntax error. MAPLE V also gave a warning about the missing semi-colon! If you forget the semi-colon, simply type it on the next line.

```
>   105/25
>   ;
```
$$21/5$$

See section 1.3 for a method for editing mistakes.

Results can be assigned to variables using the colon-equals ":=".

```
>   f:=";
```
$$f := 21/5$$

```
>  G:= -1/5;
```
$$G := -1/5$$

```
>  f+g;
```
$$21/5 + g$$

```
>  #Observe that Maple is case sensitive.
>  f+G;
```
$$4$$

Note that comment lines begin with #.

1.3   Editing mistakes

MAPLE V has built-in editing facilities. On most platforms, lines of input can be edited using the arrow keys and the mouse. Cutting and pasting is also possible with the mouse. In the Windows version, you can select input by highlighting with the mouse, then you can copy, cut, and paste by using *control-c*, *x*, and *v* as usual. In the command-line (or tty) version, MAPLE V has two built-in editors: *emacs* and *vi*. Use the help command **?editing** for more information.

```
>  105/25
>  105/25;
syntax error, unexpected number
```

Just click the mouse after 105/25, enter a semicolon, and press enter.

```
>  105/25;
```
$$21/5$$

The *vi* editor can be invoked using the *Esc* key.

1.4 Help

In MAPLE V (Release 4) there is a fabulous interactive help facility. Click on Help and select Full Text Search ... A little window should appear. In the Words box, type `floating point` then click on Search. A search is then made of the interactive help manual. A list of topics should appear in the Matching Topics box. Select `evalf` with the mouse, then click on Apply. A help window should appear with information on the `evalf` command.

If you know the name of a command, then you can select Topic Search ... in the Help window. Help can also be accessed directly from the worksheet. Try

```
> ?evalf
```

The `evalf` help window should appear. In the tty version, this information will appear below the cursor.

Now try selecting Balloon Help in the Help menu. Next move the cursor onto a button and a little bubble should appear giving a brief description. Keep this option until you are familiar with the buttons and menus.

The command `?index` provides a list of categories: expression, function, misc., etc. For instance, `?index[function]` gives a list of MAPLE

V's standard library functions. For more information on navigating through the worksheet environment, see ?worksheet[how to].

1.5 Quitting MAPLE V

If you are done with your MAPLE V session click on 💾. The *Save As* window should appear. In the File Name box type ch1.mws and click on OK. Your worksheet has now been saved. To quit maple, go to the File menu and select Exit. Later you can re-open your worksheet by clicking on 📂.

In the tty version, the easiest way to quit a Maple session is to use quit.

```
>   quit
```

2.   MAPLE V AS A CALCULATOR

2.1   Exact arithmetic and basic functions

As we noted in Section 1.2, MAPLE V does exact arithmetic. Also, MAPLE V does integer arithmetic to infinite precision. Try the following examples.

```
>   2/3 + 3/5;
```
$$\frac{19}{15}$$

```
>   7 - 11/15;
```
$$\frac{94}{15}$$

```
>   12^20;
```

$$3833759992447475122176$$

The basic arithmetic operations in MAPLE V are

+	addition
−	subtraction
*	multiplication
^or **	exponentiation
/	division

MAPLE V also has the basic mathematical functions (and much more) that are available on a scientific calculator.

abs(x)	absolute value	$\|x\|$
sqrt(x)	square root	$\sqrt{x}$
n!	factorial	
sin(x)	sine	
cos(x)	cosine	
tan(x)	tan	
sec(x)	secant	
csc(x)	cosecant	
cot(x)	cotangent	
log(x)	natural logarithm	
also ln(x)		
exp(x)	exponential function	e^x
sinh(x)	hyperbolic sine	
cosh(x)	hyperbolic cosine	
tanh(x)	hyperbolic tan	

MAPLE V has many other built-in mathematical functions. For instance, it has the inverse trig functions (**arcsin**, **arccos**, etc.), the Bessel functions (**BesselI**), the Riemann zeta function (**Zeta**), the gamma function (**GAMMA**), and the complete and incomplete elliptic integrals (**EllipticE**). For a complete listing, see **?index[functions]**.

2.2   Floating-point arithmetic

MAPLE V can do floating-point calculation to any required precision. This is done using **evalf**.

```
>   tan(Pi/5);
```

$$\sqrt{5 - 2\sqrt{5}}$$

```
>   evalf(");
```

$$0.7265425273$$

Notice that **evalf** found $\tan(\pi/5)$ to 10 decimal places which is the default. Also, note that in MAPLE V, π is represented by **Pi**. There are two ways to change the default and increase the number of decimal places.

```
>   E := exp(1): evalf(E,20);
```

$$2.7182818284590452354$$

```
> Digits := 30;
```

$$Digits := 30$$

```
> evalf(E);
```

$$2.71828182845904523536028747135$$

Here E is the mathematical constant e, which we have represented in MAPLE V by `exp(1)`. We found e to 20 digits using `evalf(E,20)`. The other method is to use the global variable `Digits` (whose default value is 10). After assigning `Digits :=` 30, we found e correct to 30 digits simply by calling `evalf(E)`. We reset the default and calculated $\sin(\pi/6)$.

```
> Digits := 10:
> evalf(sin(Pi/6));
```

$$0.5000000000$$

```
> convert(",rational);
```

$$1/2$$

Notice that after we found the decimal approximation, we were able to convert it into an exact rational using `convert(",rational)`. The `convert` function is used to convert expressions from one type to another. More on the **convert** function is to be found in Section 4.6. The interested reader can find more using **?convert**. Below is a table of MAPLE V's built-in mathematical constants.

`Catalan`	Catalan's constant (about .9159655942)
`gamma`	Euler's constant (about 0.5772156649)
`I`	complex number i $(i^2 = -1)$
`Pi`	π (about 3.141592654)

3. HIGH SCHOOL ALGEBRA

3.1 Polynomials and rational functions

3.1.1 Factoring a polynomial

MAPLE V can do high school algebra. It can manipulate polynomials and rational functions of one or more variables quite easily.

```
>  p := x^2+5*x+6;
```

$$p := x^2 + 5x + 6$$

```
>  factor(p);
```

$$(x + 3)(x + 2)$$

```
>  b := 1 - q^7 - q^8 - q^9 + q^15 + q^16
```

11

```
      + q^17 - q^24;
```

$$b := 1 - q^7 - q^8 - q^9 + q^{15} + q^{16} + q^{17} - q^{24}$$

```
>   factor(b);
```

$$-(q+1)(q^2+1)(q^2+q+1)(q^6+q^3+1)(q^4+1)$$
$$(q^6+q^5+q^4+q^3+q^2+q+1)(q-1)^3$$

To factor a polynomial or rational function we use **factor**. We let $p = x^2 + 5x + 6$ and found the factorization using **factor(p)**. This could have easily been done by hand. Factoring $b = 1 - q^7 - q^8 - q^9 + q^{15} + q^{16} + q^{17} - q^{24}$ is not so easy, but child's play for MAPLE V.

3.1.2   Expanding an expression

To expand a polynomial use **expand**. The command **combine** is also useful for expanding certain expressions.

```
>   p := (x+2)*(x+3);
```

$$p := (x+2)(x+3)$$

```
>   expand(");
```

$$x^2 + 5x + 6$$

```
>   (1-q^8)*(1-q^7)*(1-q^6);
```

$$(1-q^8)(1-q^7)(1-q^6)$$

```
>   expand(");
```

$$1 - q^7 - q^8 - q^9 + q^{15} + q^{16} + q^{17} - q^{24}$$

```
>   sqrt(x+2)*sqrt(x+3);
```

$$\sqrt{x+2}\,\sqrt{x+3}$$

```
>   expand(");
```

$$(x+2)^{1/2}(x+3)^{1/2}$$

```
>   combine(");
```

$$\sqrt{x^2 + 5x + 6}$$

Notice we were not able to expand the expression $(x+2)^{1/2}(x+3)^{1/2}$ with **expand** and had to use **combine** instead.

3.1.3   Collecting like terms

The **collect** function is useful when looking at a polynomial in more than one variable.

```
>   (x+y+1)*(x-y+1)*(x-y-1);
```

$$(x+y+1)(x-y+1)(x-y-1)$$

```
>   p := expand(");
```

$$p := x^3 - x^2y + x^2 - 2xy - x - y^2x + y^3 + y^2 - y - 1$$

```
>  collect(p,x);
```

$$x^3 + (1 - y)\,x^2 + \left(-1 - y^2 - 2\,y\right)x - y - 1 + y^3 + y^2$$

We let $p = (x + y + 1)(x - y + 1)(x - y - 1) = x^3 - x^2y + x^2 - 2\,xy - x - y^2x + y^3 + y^2 - y - 1$. We used `collect(p,x)` to write p as a polynomial in x with coefficients that were polynomials in the remaining variable y. Similarly, try `collect(p,y)` to get p as a polynomial in y.

3.1.4 Simplifying an expression

The first thing you should try when presented with a complicated expression is `simplify`.

```
>  3*4^(1/2)+5;
```

$$3\sqrt{4} + 5$$

```
>  simplify(");
```

$$11$$

```
>  x^2;
```

$$x^2$$

```
>  "^(1/2);
```

$$\sqrt{x^2}$$

```
>  simplify(");
```

$$\mathbf{csgn}(x)\,x$$

14

Notice we were able to simplify $3\sqrt{4} + 5$ to 11. Of course, the value of $(x^2)^{1/2}$ depends on the sign of x. Here **csgn** is a function that returns 1 if x is positive and -1 otherwise. If we know that $x > 0$, we can use **assume** to do further simplification ($x^\sim$ replaces x).

```
>   y:=((x-2)^2)^(1/2);
```

$$y := \sqrt{(x-2)^2}$$

```
>   assume(x>2);
>   simplify(y);
```

$$x^\sim - 2$$

3.1.5 Simplifying radicals

To simplify expressions using radicals, we can use **simplify** and **radsimp**. First, we remove the assumption on x

```
>   x := 'x';
```

$$x := x$$

This restores x to its original status. See Section 3.1.9.

```
>   y := x^3 + 3*x^2 + 3*x + 1;
```

$$y := x^3 + 3x^2 + 3x + 1$$

15

```
>    simplify(y^(1/3));
```

$$((1+x)^3)^{1/3}$$

```
>    radsimp(y^(1/3));
```

$$1+x$$

```
>    assume(x>-1);
>    simplify(y^(1/3));
```

$$1+x^\sim$$

```
>    assume(x<-1);
>    simplify(y^(1/3));
```

$$-1/2\,(x^\sim + 1)\,(1 + I\,3^{1/2})$$

```
>    x := 'x':
```

Notice that simplify recognized y as a cube but failed to simplify $y^{1/3}$. The command radsimp, on the other hand, was able to simplify $y^{1/3}$ to $1+x$. If assumptions are given for x, then simplify is able to simplify the radical further. However, it should be noted that the value of the cube root depends on these assumptions so care needs to be taken.

A cute MAPLE V command is rationalize. However, before using it, we must first use readlib to read it into memory. Most of MAPLE V's functions are automatically loaded when a maple session is started. Other functions in various packages (see Chapter 11) are read in using `with`. In addition, there are some functions that are only read in using `readlib`.

```
>   readlib(rationalize):
```

```
>   1/(1+sqrt(2));
```

$$\frac{1}{\sqrt{2}+1}$$

```
>   rationalize(");
```

$$\sqrt{2}-1$$

```
>   (1-2^(2/3))/(1+2^(1/3));
```

$$\frac{1-2^{2/3}}{1+2^{1/3}}$$

```
>   rationalize(");
```

$$-2^{1/3}+1$$

```
>   y:= z/(1 + sqrt(x));
```

$$y := \frac{z}{1+\sqrt{x}}$$

```
> rationalize(y);
```

$$\frac{z\left(-1+\sqrt{x}\right)}{-1+x}$$

Notice that **rationalize** does a great job rationalizing a denominator not only for expressions involving square roots but for more complicated radicals as well. It can also handle symbolic expressions.

3.1.6   Simplifying rational functions

To simplify a rational function (i.e., a function that can be written as a quotient of two polynomials) we use the command **normal**. This has the effect of cancelling any common factors between numerator and denominator. First we restore x and y's variable status.

```
> y:='y':  z:='z':
> a:= (x-y-z)*(x+y+z);
```

$$a := (x - y - z)(x + y + z);$$

```
> b :=(x^2-2*x*y-2*x*z+y^2+2*y*z+z^2)
    *(x^2-x*y+x*z-y*z);
```

$$b := (x^2 - 2xy - 2xz + y^2 + 2yz + z^2)(x^2 - xy + xz - yz)$$

```
> c:=a/b: normal(c);
```

$$-\frac{(x + y + z)}{(x^2 - yx + xz - yz)(-x + y + z)}$$

```
>  simplify(c);
```

$$-\frac{(x+y+z)}{(x^2-yx+xz-yz)(-x+y+z)}$$

```
>  factor(c);
```

$$-\frac{(x+y+z)}{(x-y)(x+z)(-x+y+z)}$$

Observe that **normal** and **simplify** had the same effect on the rational function c. We use **normal** for rational functions if we can do without the more expensive **simplify**. Also, we could have used **factor** to simplify c and get it into a nice form. It should be noted that **normal** is able to do this simplication without factoring, which is more expensive.

Some useful functions for manipulating rational functions are: **numer**, **denom**, **rem**, and **quo**. We let c be as above.

```
>  numer(c);
```

$$-(-x+y+z)(x+y+z)$$

```
>  denom(c);
```

$$(x^2-2xy-2xz+y^2+2yz+z^2)(x^2-xy+xz-yz)$$

```
>  factor(");
```

$$(-x+y+z)^2(x-y)(x+z)$$

19

The functions **numer** and **denom** select the numerator and denominator, respectively, of a rational function. After factorizing the denominator of c, we see that there was simplification because of the common factor $(-x + y + z)$.

The functions **quo** and **rem** give the quotient and remainder upon polynomial division.

```
>   a:= 2*x^3+x^2+12;
```

$$a := 2x^3 + x^2 + 12$$

```
>   b := x^2 - 4;
```

$$b := x^2 - 4$$

```
>   q := quo(a,b,x);
```

$$q := 2x + 1$$

```
>   r := rem(a,b,x);
```

$$r := 16 + 8x$$

```
>   expand( a - (b*q + r) );
```

$$0$$

The command **quo(a,b,x)** gives the quotient q when a is divided by b as polynomials in x. The command `rem(a,b,x)` gives the remainder r so that

$$a = bq + r,$$

and the degree of r (as a polynomial in x) is less then the degree of b.

3.1.7   Coefficients of a polynomial

In Section 3.1.3 the `collect` command was introduced to view polynomials. Two other useful functions are `coeff` and `degree`. Let p be as before.

```
>  p:= (x+y+1)*(x-y+1)*(x-y-1):
>  q := expand("):
>  coeff(q,x,2);
```

$$-y + 1$$

```
>  coeff(p,x,2);
```
$$0$$

```
>  degree(q,x);
```
$$3$$

The command `coeff(q,x,2)` found the coefficient of x^2 in the polynomial q. The command `degree(q,x)` gave the degree of q as a polynomial in x. Observe also that when `coeff` was applied to the unexpanded form p, an "incorrect" value of 0 was returned. Be careful.

3.1.8 Substituting into an expression

We may substitute into an expression using the command **subs**.

```
>  p := (x+y+z)(x-y+z)(x-y-z);
```

$$p := (x + y + z)(x - y + z)(x - y - z)$$

```
>  subs(x=1,p);
```

$$(1 + y + z)(1 - y + z)(1 - y - z)$$

To substitute $x = 1$ into p, we used the command **subs(x=1,p)**. Try substituting $x = 1$ and $y = 2$ into p using the command **subs(x=1,y=2,p)**.

3.1.9 Restoring variable status

In the last section we saw how **subs** is used to do substitution. There is another way to do this. We let p be as Section 3.1.8.

```
>  p;
>  x:=1:   y:=2:
>  p;
```

$$(3 + z)(-1 + z)(-1 - z)$$

We are able to do the subsitution by assigning $x :=$ 1 and $y := 2$. However, now p has changed. There is a way to restore x and y's variable status.

```
> x := 'x':   y := 'y':
> p;
```

$$(x + y + z)(x - y + z)(x - y - z)$$

The assignments x := 'x' and y := 'y' restored x and y to their variable status. It is neat that p was also restored to its original status.

3.2   Equations

3.2.1   Left- and right-hand sides

To assign a value to a variable we use :=. The symbol = has a different meaning and is reserved for equations.

```
> eqn := x^2 - x = 1;
```

$$eqn := x^2 - x = 1$$

```
> R := solve(eqn,x);
```

$$R := 1/2\sqrt{5} + 1/2, \quad 1/2 - 1/2\,5^{1/2}$$

```
> simplify(R[1]*R[2]);
```

$$-1/4\,(\sqrt{5} + 1)\,(\sqrt{5} - 1)$$

```
> expand(");
```

$$-1$$

23

We assigned to equation $x^2 - x = 1$ the name *eqn*. We solved the equation for x by typing `solve(eqn,x)`. We named the list of solutions R. The two solutions were $R[1]$ and $R[2]$. In this way we can manipulate the solutions. Observe that we computed the product of the roots to be -1 as expected.

The left and right sides of an equation can be manipulated using `lhs` and `rhs`.

```
>   eqn;
```
$$x^2 - x = 1$$

```
>   lhs(eqn);
```
$$x^2 - x$$

```
>   subs(x=R[1],lhs(eqn));
```
$$(1/2 + 1/2\sqrt{5})^2 - 1/2\sqrt{5} - 1/2$$

```
>   expand(");
```
$$1$$

The command `lhs(eqn)` gave us the left side of the equation. Then we were able to substitute $x = R[1]$ (the first root) into the left side of the equation, which simplified to 1 (as expected) using `expand`.

3.2.2   Finding exact solutions

MAPLE V has the capability for solving systems of equations.

```
>  eqn1:= x^3+a*x=14; eqn2 := a^2-x=7;
```

$$eqn1 := x^3 + ax = 14$$
$$eqn2 := a^2 - x = 7$$

```
>  solve({eqn1,eqn2},{x,a});
```

$\{a = 3, x = 2\},$
$\{a = \text{RootOf}(\_Z^5 + 3\_Z^4 - 12\_Z^3 - 35\_Z^2 + 42\_Z$
$\quad + 119), x = \text{RootOf}(\_Z^5 + 3\_Z^4 - 12\_Z^3$
$\quad - 35\_Z^2 + 42\_Z + 119)^2 - 7\}$

The syntax for solving systems of equations is
solve(S,X) where S is a set of equations and X is
the required set of variables. Observe that MAPLE
V was able to find the solution $x = 2$, $a = 3$. It
also found that $a = z$, $x = z^2 - 7$ are solutions
where z is any root of the following polynomial
equation:

$$Z^5 + 3Z^4 - 12Z^3 - 35Z^2 + 42Z + 119 = 0.$$

As in the previous section, we may manipulate solutions. We select the first set of solutions and
substitute them into the first equation.

```
>  "[1];
```

$$\{a = 3, x = 2\}$$

```
>  subs(",eqn1);
```

$$14 = 14$$

3.2.3   Finding approximate solutions

In the last section we came upon the following quintic

$$Z^5 + 3Z^4 - 12Z^3 - 35Z^2 + 42Z + 119 = 0.$$

Although naturally enough MAPLE V is unable to find an exact explicit solution, it is able to find approximate solutions using **fsolve**.

```
>  polyeqn := Z^5+3*Z^4-12*Z^3-35*Z^2
     +42*Z+119=0:
>  a1 := fsolve(polyeqn,Z);
```

$$a1 := -3.136896207$$

```
>  x1:= a1^2 -7;
```

$$x1 := 2.840117813$$

```
>  subs({x=x1,a=a1},{eqn1,eqn2});
```

$$\{14.00000003 = 14, \ 7.0000000000 = 7\}$$

We used the command **fsolve(polyeqn,Z)** to find the approximate solution $Z \approx -3.136896207$. This implied that $a = -3.136896207$ and $x = a^2 - 7 = 2.840117813$ are approximate solutions to our system of equations in the previous section. We were able to check this using **subs**.

3.2.4   Assigning solutions

Once an equation or system of equations has been solved, we can use **assign** to assign a particular solution to the variable(s). We use the example given in Section 3.2.2.

```
>   solve({x^3+a*x=14,a^2-x=7},{a,x}):
>   "[1];
```
$$\{a = 3, x = 2\}$$

```
>   assign(");
>   a; x;
```

$$3$$
$$2$$

3.3   Fun with integers

3.3.1   Complete integer factorization

The command **ifactor** gives the prime factorization of an integer.

```
>   2^(2^5)+1;
```

$$4294967297$$

```
>  ifactor(");
```

$$(641) (6700417)$$

```
>  ifactor(5003266235067621177579);
```

$$(3)^2 (13) (31)^3 (67) (139) (320057) (481577)$$

3.3.2   Quotient and remainder

The integer analogs of quo and rem, the functions for finding the quotient and the remainder in polynomial division, are the functions iquo and irem. They work in the same way.

```
>  a := 23;    b := 5;
```

$$a := 23$$
$$b := 5$$

```
>  q := iquo(a,b);   r := irem(a,b);
```

$$q := 4$$
$$r := 3$$

```
>  b*q+r;
```

$$23$$

We observe that if q = iquo(a,b) and r = irem(a,b), then

$$a = bq + r,$$

where $0 \leq r < b$ if a and b are positive.

Two related functions are floor and frac. floor(x) gives the greatest integer less than or equal to x and frac(x) gives the fractional part of x. Try

```
>  x := 22/7;
>  floor(x);
>  frac(x);
>  floor(-x);
>  frac(-x);
```

3.3.3   Gcd and lcm

The greatest common divisor and the lowest common multiple of a set of numbers can be found using gcd and lcm.

```
>  gcd(28743,552805);
```

$$11$$

```
>  ifactor(28743);    ifactor(552805);
```

$$(3)\,(11)\,(13)\,(67)$$
$$(5)\,(11)\,(19)\,(23)^2$$

```
>   lcm(21,35,99);
```

$$3465$$

We find that the gcd of 28743 and 552805 is 11.
This can also be seen from the prime factorizations. The lcm of 21, 35 and 99 is 3465.

3.3.4   Primes

The i-th prime can be computed with
ithprime. The function isprime tests whether a
given integer is prime or composite.

```
>   ithprime(100);
```

$$541$$

```
>   isprime(2^101-1);
```

false

```
>   7*3^10 + 10;
```

$$413353$$

```
>   isprime(");
```

true

We found that the 100th prime is 541, that $2^{101} - 1$
is composite, and that $7 \cdot 3^{10} + 10 = 413353$ is
prime.

3.3.5 Integer solutions

In Sections 3.2.1 and 3.2.2 we saw how to solve equations in MAPLE V using **solve**. The integer analog of **solve** is **isolve**. We use this function if we are only interested in integer solutions. We use the example from Section 3.2.2. Remember to restore variable status to x and a first.

```
>  x:='x':  a:='a':
>  eqn1:= x^3+a*x=14:  eqn2 := a^2-x=7:
>  isolve({eqn1,eqn2},{x,a});
```

$$\{a = 3, x = 2\}$$

This time we found the unique integer solution $a = 3$, $x = 2$ to the given system of equations.

4.   DATA TYPES

4.1   Sequences

In MAPLE V *sequences* take the form

$$expr1, \; expr2, \; expr3, \; \ldots, \; exprn.$$

```
>  x := 1,2,3;
```

$$x := 1, 2, 3$$

```
>  y := 4,5,6;
```

$$y := 4, 5, 6;$$

```
>  x,y;
```
$$1, 2, 3, 4, 5, 6$$

We observe that in MAPLE V, **x,y** concatenates the two sequences x and y. There are two important functions used to construct sequences: **seq** and the repetition operator **$**.

```
>  f:='f':   seq(f(i), i=1..6);
```
$$f(1), f(2), f(3), f(4), f(5), f(6)$$

```
>  seq(i^2, i=1..5);
```
$$1, 4, 9, 16, 25$$

```
>  x:= 'x':
>  x$4;
```
$$x, x, x, x$$

In general, **seq(f(i), i=1..n)** produces the sequence
$$f(1), f(2), \ldots, f(n)$$

and **x$n** produces a sequence of length n
$$x, x, \ldots, x$$

The **op** function can be used to create sequences.

```
>  b:='b':   c:='c':
>  L := a+b+2*c+3*d;
```
$$L := a + b + 2c + 3d$$

```
>   op(");
```
$$a, b, 2c, 3d$$

op(expr) produces a sequence whose elements are the operands in expr.

```
>   nops(L);
```
$$4$$

```
>   op(3,L);
```
$$2c$$

nops(expr) gives the length of the sequence op(expr) and op(j,expr) gives the j-th term in the sequence op(expr).

If s is a sequence, then the j-th term of the sequence is $s[j]$.

```
>   s := 1, 8, 27, 64, 125;
```
$$s := 1, 8, 27, 64, 125$$

```
>   s[3];
```
$$27$$

4.2   Sets

We have already seen the set data type in Section 3.2.2 when solving systems of equations. In MAPLE V, a *set* takes the form

$$\{expr1, expr2, expr3, \ldots, exprn\}.$$

In other words, a set has the form $\{S\}$ where S is a sequence. A set is a set in the mathematical sense — order is not important.

```
> y := 'y':   {x,y,z,y};
```

$$\{x, y, z\}$$

Observe that $\{x, y, z, y\} = \{x, y, z\}$. MAPLE V can perform the usual set operations: *union*, *intersection*, and *difference*.

```
>   a := {1,2,3,4}; b := {2,4,6,8};
```

$$a := \{1, 2, 3, 4\}$$
$$b := \{2, 4, 6, 8\}$$

```
>   a union b;
```

$$\{1, 2, 3, 4, 6, 8\}$$

```
>   a intersect b;
```

$$\{2, 4\}$$

```
>   a minus b;
```

$$\{1, 3\}$$

We can also determine whether a given expression is an element of a set using the function **member**.

```
>  member(2,a);
```
$$true$$

```
>  member(5,a);
```

$$false$$

```
>  a[3];
```
$$3$$

So member(x,A) returns the value true if x is an element of A and false otherwise. Also, the j-th element of the set A is A[j].

4.3   Lists

In MAPLE V, a *list* takes the form

$$[expr1, \; expr2, \; expr3, \; \ldots, \; exprn].$$

Here order is important.

```
>  a:='a':  b:='b':
>  L1 := [x,y,z,y]; L2 := [a,b,c];
```

$$L1 := [x, \; y, \; z, \; y]$$
$$L2 := [u, \; b, \; c]$$

```
>  L := [op(L1),op(L2)];
```

$$L := [x, \; y, \; z, \; y, \; a, \; b, \; c]$$

```
>  L[5];
```

$$a$$

We observe that the lists $L1$ and $L2$ can be concatenated by the command `[op(L1),op(L2)]` and that `L[j]` gives the j-th item in the list L. Lists can be created from sequences:

```
>  s := seq( i/(i+1), i=1..6);
```

$$s := 1/2,\ 2/3,\ 3/4,\ 4/5,\ 5/6,\ 6/7$$

```
>  M := [s];
```

$$M := [1/2,\ 2/3,\ 3/4,\ 4/5,\ 5/6,\ 6/7]$$

```
>  M[2..5];
```

$$[2/3,\ 3/4,\ 4/5,\ 5/6]$$

So, `M[i..j]` gives the i-th through j-th elements of the list M.

4.4 Tables

In MAPLE V, a *table* is an array of expressions whose indexing set is not necessarily a set of integers. Sounds bizarre? — let's look at some examples. Tables are created by the `table` function.

```
>  T := table([a,b]);
```

$$T := \text{table}([$$
$$1 = a$$
$$2 = b$$
$$])$$

```
>  T[2];
```

$$b$$

So, if L is a list, then `table(L)` converts L into a table. The j-th element of this table T is given by `T[j]`. Try

```
>  S := table([(1)=A,(3)=B+C,(5)=A*B*C]);
>  S[3];
>  S;
>  op(S);
```

For the table S, the indexing set is $\{1, 3, 5\}$ and thus does not necessarily have to be a set of consecutive integers. See `?table` for more bizarre examples. In your session you should have found that S did not return the table, but that `op(S)` did.

4.5   Arrays

In MAPLE V, an *array* is a special kind of a table. It most resembles a matrix. Let's look at some examples.

```
> A := array(1..2,1..3,[ ]);
```

$$A := \mathrm{array}(1..2, 1..3)$$

```
>  op(A);
```

$$\begin{bmatrix} ?_{1,1} & ?_{1,2} & ?_{1,3} \\ ?_{2,1} & ?_{2,2} & ?_{2,3} \end{bmatrix}$$

```
> B := array(1..2,1..2,1..2,[ ]);
```

$$B := \mathrm{array}(1..2, 1..2, 1..2)$$

```
>  op(B);
```

$$\begin{aligned}
\mathrm{array}(1..2, 1..2, 1..2, [\\
(1,1,1) =& ?_{1,1,1} \\
(1,1,2) =& ?_{1,1,2} \\
(1,2,2) =& ?_{1,2,2} \\
(2,1,1) =& ?_{2,1,1} \\
(2,1,2) =& ?_{2,1,2} \\
(2,2,1) =& ?_{2,2,1} \\
(2,2,2) =& ?_{2,2,2} \\
])
\end{aligned}$$

We see that the array A corresponds to a 2×3 matrix. The array B corresponds to $2 \times 2 \times 2$ matrix or, if you like, a table with indexing set

$$\{(1,1,1), (1,1,2), \ldots, (2,2,2)\}.$$

```
```

We can insert entries into an array by using the subscripts (or indices).

```
> C:=array(1..2,1..2):
> C[1,1]:=1:  C[1,2]:=2:  C[2,1]:=3:
  C[2,2]:=7:
> op(C);
```

$$\begin{bmatrix} 1 & 2 \\ 3 & 7 \end{bmatrix}$$

An alternative method is given below.

```
> F:=array(1..2,1..3,[[1,2,3],[5,9,7]]);
```

$$F := \begin{bmatrix} 1 & 2 & 3 \\ 5 & 9 & 7 \end{bmatrix}$$

4.6   Data conversions

The function **type** checks the data type of an object.

```
> A := {1,2,3}:
> s := 1,2,3:
> L := [1,2,3]:
> T := table([1,2,3]):
> M := array(1..3,[1,2,3]):
> type(L,list);
```

true

```
>  type(T,set);
```

$$\text{false}$$

The function **convert** can be used to convert from one data type to the other.

```
>  convert(A,list);
```

$$[1, 2, 3]$$

```
>  convert(L,set);
```

$$\{1, 2, 3\}$$

Try using the function **whattype**. See **?whattype** for help.

5.   CALCULUS

5.1   Defining functions

To enter the function $f(x) = x^2 - 3x + 5$, type

```
>  f:= x -> x^2 - 3*x + 5;
```

$$f := x \rightarrow x^2 - 3x + 5$$

The arrow symbol is entered by typing the *minus* key – immediately followed by the *greater than* key >. We compute $f(2)$.

```
> f(2);
```

Thus, in MAPLE V the syntax for creating a function $f(x)$ is `f := x -> expr`, where `expr` is some expression involving x. Functions in more than one variable are defined in the same way.

```
> g := (x,y) -> x*y/(1+x^2+y^2);
```

$$g := (x, y) \rightarrow \frac{xy}{1 + x^2 + y^2}$$

We defined the function

$$g(x, y) = \frac{xy}{1 + x^2 + y^2}.$$

Try simplifying $g(\sin t, \cos t)$

```
> g(sin(t),cos(t));
> simplify(");
```

To convert an expression into a function, we use the **unapply** function.

```
> q := Z^5+3*Z^4-12*Z^3-35*Z^2
      +42*Z+119:
> h := unapply(q,Z);
```

$$h := Z \rightarrow Z^5 + 3Z^4 - 12Z^3 - 35Z^2 + 42Z + 119$$

In Sections 3.2 and 3.3 we came across the quintic polynomial q above. Here q is an expression involving Z. To convert q into the function $h(Z)$, we used the command `unapply(q,Z)`. Now we are free to play with the function h.

```
>  H := x -> evalf( h(x), 4):
```

$$H := x \to \text{evalf}(h(x), 4)$$

```
>  X := [seq(evalf(-4+i/10,4),i=0..10)];
```

$X := [-4., -3.900, -3.800, -3.700, -3.600,$
$\quad -3.500, -3.400, -3.300, -3.200, -3.100, -3.]$

```
>  Y := map(H,X);
```

$Y := [-97., -73.7, -54.5, -39.0, -26.6, -17.1,$
$\quad -10.4, -5.1, -1.4, .7, 2.]$

The function $H(x)$ computes $h(x)$ to 4 digits. Then we used `seq` and `map` to produce the lists X and Y which give a table of x and y values for the function $y = h(x)$.

5.2   Composition of functions

In MAPLE V, `@` is the function composition operator. If f and g are functions, then the composition of f and g, $f \circ g(x) = f(g(x))$, is given by `(f@g)(x)`.

```
>  (sin@cos)(x);
```

$$\sin(\cos(x));$$

```
> f := x -> x^2:
> g := x -> sqrt(1-x):
> (f@g)(x);
```
$$1 - x$$

```
> (g@f)(x);
```
$$\sqrt{1 - x^2}$$

`@@` gives repeated composition so that `(f@@2)(x)` gives $f(f(x))$ and `(f@@3)(x)` gives $f(f(f(x)))$. For certain functions known to MAPLE V, `f@@(-1)(x)` gives the inverse function $f^{-1}(x)$.

5.3 Summation and product

In MAPLE V, the syntax for the sum

$$\sum_{i=1}^{n} f(i) = f(1) + f(2) + \cdots + f(n)$$

is `Sum(f(i),i=1..n)` and `sum(f(i),i=1..n)`.

```
> f := 'f':
> Sum(f(i),i=1..n);
```

$$\sum_{i=1}^{n} f(i)$$

```
> Sum(i^2,i=1..10);
```

$$\sum_{i=1}^{10} i^2$$

43

```
>  sum(i^2,i=1..10);
```

$$385$$

Notice that the difference between **sum** and **Sum** is that in **sum**, the sum is evaluated, but that in **Sum**, it is not. It is recommended that one get into the habit of using **Sum** to first check for typos and then use **value** to evaluate the sum. In our previous session we found

$$\sum_{i=1}^{10} i^2 = 1 + 4 + 9 + \cdots + 100 = 385.$$

This time we will use **Sum** and **value**.

```
>  Sum(i^2,i=1..10);
```

$$\sum_{i=1}^{10} i^2$$

```
>  value(");
```

$$385$$

```
>  sum(i^2,i=1..n);
```

$$1/3\,(n+1)^3 - 1/2\,(n+1)^2 + 1/6\,n + 1/6$$

```
>  factor(");
```

$$\frac{1}{6}n(n+1)(2n+1)$$

Notice that MAPLE V knows certain summation formulas such as

$$\sum_{i=1}^{n} i^2 = \frac{1}{6}n(n+1)(2n+1).$$

In MAPLE V, the syntax for the product

$$\prod_{i=1}^{n} f(i) = f(1) \cdot f(2) \cdots f(n)$$

is Product(f(i),i=1..n).

```
>  f := 'f': q := 'q':
>  Product(f(i),i=1..n);
```

$$\prod_{i=1}^{n} f(i)$$

```
>  Product(1-q^i,i=1..5);
```

$$\prod_{i=1}^{5} 1 - q^i$$

45

```
>   value(");
```

$$(1 - q)(1 - q^2)(1 - q^3)(1 - q^4)(1 - q^5)$$

```
>   expand(");
```

$$-q^{15} + q^{14} + q^{13} - q^{10} - q^9 - q^8 + q^7 + q^6 + q^5 - q^2 - q + 1$$

As with **sum** and **Sum**, for **product**, the product is evaluated, but with **Product**, it is not. Note that we could've evaluated the product $\prod_{i=1}^{5} 1 - q^i$ using **product(1-q^i,i=1..5)**.

A common problem with **sum** and **product** is the following.

```
>   i:=2;
```

$$i := 2$$

```
>   sum(i^3,i=1..5);
Error, (in sum) summation variable
previously assigned, second argument
evaluates to, 2=1 ..  5
```

The problem occurred in **sum** since i had already been assigned the value 2. There are two ways around this problem. One way is to restore the variable status of i by typing i := 'i'. The second way is to replace i by 'i' in the sum.

```
>   sum('i'^3,'i'=1..5);
```

225

5.4 Limits

Naturally, there are two forms of the MAPLE V limit function: `Limit` and `limit`. These are analogous to sum and Sum, etc.

The syntax for computing the limit of $f(x)$ as $x \to a$ is `Limit(f(x), x=a); value(")`. The `Limit` command displays the limit so that it can be checked for typos and then the **value** command computes the limit. To compute the limit

$$\lim_{x \to 2} \frac{x^2 - 4}{x - 2}$$

we type

```
>   Limit((x^2-4)/(x-2),x=2); value(");
```

$$\lim_{x \to 2} \frac{x^2 - 4}{x - 2}$$

$$4$$

Thus, we see that

$$\lim_{x \to 2} \frac{x^2 - 4}{x - 2} = 4,$$

which can be verified easily with paper and pencil. Alternatively, we could've found the limit in one step by typing `limit((x^2-4)/(x-2),x=2)`.

Left and right limits can also be calculated as well as limits where x approaches infinity. Try

```
>   f:=(x^2-4)/(x^2-5*x+6);
>   Limit(f,x=3,right); value(");
>   Limit(f,x=infinity); value(");
```

5.5   Differentiation

MAPLE V can easily find the derivatives of functions of one or several variables. The syntax for differentiating $f(x)$ is `diff(f(x),x)`.

```
>   f := sqrt(1 - x^2);
```

$$f := \sqrt{1 - x^2}$$

```
>   diff(f,x);
```

$$-\frac{x}{\sqrt{1 - x^2}}$$

```
>   g := z -> z^2*exp(z) + sin(log(z)):
>   diff(g(x),x);
```

$$2x\,e^x + x^2\,e^x + \frac{\cos(\ln(x))}{x}$$

The second derivative is given by typing `diff(f(x),x,x)`. For the n-th derivative, use `diff(f(x),x$n)`. Use MAPLE V to show that

$$\frac{d^5 \tan x}{dx^5} = 136 \tan^2 x + 240 \tan^4 x$$
$$+ 120 \tan^6 x + 16.$$

In MAPLE V, partial derivatives are computed using `diff`.

```
>   z := exp(x*y)(1+sqrt(x^2+3*y^2-x));
```

$$z := e^{xy}\left(1 + \sqrt{x^2 + 3y^2 - x}\right)$$

```
>   diff(z,x);
```

$$ye^{xy}\left(1 + \sqrt{x^2 + 3y^2 - x}\right) + \frac{e^{xy}(2x - 1)}{2\sqrt{x^2 + 3y^2 - x}}$$

```
>   normal(diff(z,x,y)-diff(z,y,x));
```

$$0$$

The syntax for $\frac{\partial z}{\partial x}$ is `diff(z,x)` and for $\frac{\partial^2 z}{\partial y \partial x}$ is `diff(z,x,y)`. For

$$z = e^{xy}\left(1 + \sqrt{x^2 + 3y^2 - x}\right)$$

we found that

$$\frac{\partial z}{\partial x} = ye^{xy}\left(1 + \sqrt{x^2 + 3y^2 - x}\right)$$
$$+ \frac{e^{xy}(2x - 1)}{2\sqrt{x^2 + 3y^2 - x}},$$

and

$$\frac{\partial^2 z}{\partial y \partial x} = \frac{\partial^2 z}{\partial x \partial x}.$$

49

MAPLE V also has the differential operator D. If f is a differentiable function of one variable, then Df is the derivative f'. We calculate $g'(x)$ for our function g above.

```
>  g := z -> z^2*exp(z) + sin(z);
```

$$g := z \to z^2\, e^z + \sin(z)$$

```
>  D(g);
```

$$z \to 2z\, e^z + z^2 e^z + \cos(z)$$

5.6   Extrema

MAPLE V is able to find the minimum and maximum values of certain functions of one or several variables with zero or more constraints. There are three possible approaches: (1) using the built-in functions **maximize** and **minimize**, (2) using the miscellaneous library function **extrema**, and (3) using the *simplex* package (for linear functions). Here we will describe (1) and (2). See **?simplex** for (3).

The functions **maximize** and **minimize** can find the maximum and minimum values of a function of one or several variables. There is also an option for restricting some of the variables to certain intervals. It is advised that this facility be used with care and only with algebraic functions – not the transcendental functions such as `exp`, `sin`, `cos`, etc.

We can find the maximum value of the function $f(x)$ by typing `maximize(f(x))`. The command `maximize(f(x), {x},{x=a..b})` gives the maximum of the function with x restricted to the interval $[a, b]$.

```
>   maximize(sin(x)+cos(x));
```

$$\sqrt{2}$$

```
>   maximize(x^2-5*x+1,{x},{x=0..3});
```

$$1$$

```
>   maximize(sin(x),{x},{x=0..1});
```

$$1$$

We found that the maximum value of $\sin x + \cos x$ is $\sqrt{2}$. For $0 \leq x \leq 3$, the maximum value of $x^2 - 5x + 1$ was found to be 1. However, MAPLE V incorrectly computed the maximum of $\sin x$ (for $0 \leq x \leq 1$) to be 1. The function $\sin x$ is increasing on $[0, 1]$ so the actual maximum value is $\sin 1 \approx 0.841$. We hope this bug will be fixed.

To find the minimum value of a function, use the command `minimize` whose syntax is analogous to that of `maximize`. MAPLE V can also handle functions of more than one variable.

```
>   minimize(x^2+y^2,{x,y});
```

$$0$$

```
>   minimize(x^2+y^2,x);
```

$$y^2$$

We found the minimum value of $x^2 + y^2$ to be 0. The function `minimize(x^2 + y^2,x)` found the minimum value of the function $x^2 + y^2$, considered as a function of x with y fixed.

The second method involves using the misc library function `extrema`, so we must first load the desired function with `readlib(extrema)`. The function `extrema` is able to find the minimum and maximum values of algebraic functions of one or several variables, subject to 0 or more constraints. It returns a set of possible maximum and minimum values, with the option of returning a possible set of points where these values occur. The syntax for the function is `extrema(f,{g1,g2, ... ,gn},{x1,x2, ... ,xm},'s')`. Here, f is the function. The constraints are $g_1 = 0$, $g_2 = 0,\ldots,$ $g_n = 0$. $x_1, x_2, \ldots, x_m$ are the variables and s is the unevaluated variable for holding the set of possible points where the extrema occur.

```
>  readlib(extrema):
>  f := 2*x^2 + y + y^2;
```

$$f := 2x^2 + y + y^2$$

```
>  g := x^2 + y^2 - 1;
```

$$g := x^2 + y^2 - 1$$

```
>  extrema(f,{g},{x,y},'s');
```

$$\{0, 9/4\}$$

```
>  s;
```

$$\{\{x = 0, y = 1\}, \{x = 0, y = -1\}\},$$
$$\{\{y = 1/2, x = 1/2\text{RootOf}(\_Z^2 - 3)\}\}$$

```
>  simplify(subs(s[1],f));
```

$$0$$

```
>  simplify(subs(s[2],f));
```

$$2$$

```
>  simplify(subs(s[3],f));
```

$$9/4$$

53

By using the command **extrema(f,{g},{x,y}, 's')**, we found that the extreme values of $f(x,y) = 2x^2 + y + y^2$ (subject to the constraint $x^2 + y^2 = 1$) are 0 and 9/4. The set of possible points where the extrema occured was assigned to the variable s. Using **simplify** and **subs**, we substituted each set of points into f. In this way, we found that the minimum value 0 occurs at $x = 0, y = -1$ and the maximum value 9/4 occurs at $x = \pm\sqrt{3}/2, y = 1/2$.

5.7   Integration

If f is an expression involving x, then the syntax for finding the integral $\int_a^b f(x)\,dx$ is **int(f,x=a..b)**. For the indefinite integral we use **int(f,x)**. There are also the unevaluated forms **Int(f,x=a..b)** and **Int(f,x)**.

```
>   Int(x^2/sqrt(1-x^3),x);
```

$$\int \frac{x^2}{\sqrt{1-x^3}}\,dx$$

```
>   value(");
```

$$-2/3\,\sqrt{1-x^3}$$

```
>   Int(1/x/sqrt(x^2 - 1),x=1..2/sqrt(3));
```

$$\int_1^{2/\sqrt{3}} \frac{1}{x\sqrt{x^2-1}}\,dx$$

```
>    value(");
```

$$\frac{1}{6}\pi$$

MAPLE V easily found that

$$\int \frac{x^2}{\sqrt{1-x^3}}\, dx = -\frac{2}{3}\sqrt{1-x^3}$$

and

$$\int_1^{2/\sqrt{3}} \frac{1}{x\sqrt{x^2-1}}\, dx = \frac{\pi}{6}.$$

MAPLE V can do improper integrals and multiple integrals in the obvious way. Try finding

$$\int_0^\infty r e^{-r^2}\, dr$$

by   typing   `int(r*exp(-r^2),r=0..infinity)`.
Try evaluating the double integral

$$\int\int y \sin(2x + 3y^2)\, dx\, dy$$

by first integrating with respect to x and then with respect to y.

If MAPLE V does not know the value of a definite integral, try **evalf**.

```
>    Int(sqrt(1+x^6),x=0..1);
```

$$\int_0^1 \sqrt{1+x^6}\, dx$$

55

```
> value(");
```

$$\int_0^1 \sqrt{1 + x^6}\, dx$$

```
> evalf(");
```

$$1.064088379$$

Although MAPLE V was unable to evaluate the integral, it was able to find the approximation

$$\int_0^1 \sqrt{1 + x^6}\, dx \approx 1.064088379.$$

5.7.1   Techniques of integration

MAPLE V knows some standard techniques of integration. These are in the *student* package and are loaded with the command with(student).

5.7.1.1   Substitution

In MAPLE V, to do integration by substitution, we use the changevar command. The syntax is changevar(f(u)=h(x), *integral*, u) where *integral* is an integral in the variable x, $f(u) = h(x)$ is the substitution, and u is the new variable in the integral.

```
>   with(student):
>   Int(x^4/sqrt(1-x^10),x);
```

$$\int \frac{x^4}{\sqrt{1-x^{10}}} \, dx$$

```
>   G := value(");
```

$$G := \int \frac{x^4}{\sqrt{1-x^{10}}} \, dx$$

```
>   G2:=changevar(u=x^5,G,u);
```

$$1/5 \arcsin(u)$$

```
>   subs(u=x^5,G2);
```

$$1/5 \arcsin(x^5)$$

```
>   diff(",x);
```

$$\frac{x^4}{\sqrt{1-x^{10}}}$$

Although MAPLE V was unable to evaluate the integral at first, we were able to help it along by using **changevar** and the substitution $u = x^5$. MAPLE V was then able to evaluate the new integral. We substituted $u = x^5$ to obtain

$$\int \frac{x^4}{\sqrt{1-x^{10}}} \, dx = \sin^{-1}(x^5).$$

We then checked our answer using `diff`. Try evaluating the integral

$$\int \frac{x^7}{\sqrt{x^{16} - 1}} \, dx$$

using the substitution $u = x^8$.

5.7.1.2 Integration by parts

To do integration by parts, we use the command `intparts`. The syntax is `intparts(`*integral*`, x)` where x is the variable of integration in the *integral*.

```
>   Int(x*cos(3*x),x);
```

$$\int x \, \cos 3x \, dx$$

```
>   intparts(",x);
```

$$1/3 \, x \, \sin(3x) - \int 1/3 \, \sin(3x) \, dx$$

```
>   value(");
```

$$1/3 \, x \, \sin(3x) + 1/9 \, \cos(3x)$$

Thus MAPLE V has helped us by providing the working to evaluate the integral by parts:

$$\int x \, \cos 3x \, dx = 1/3 \, x \, \sin 3x - \int 1/3 \, \sin 3x \, dx$$

$$= 1/3 \, x \, \sin 3x + 1/9 \, \cos 3x.$$

5.7.1.3 Partial fractions

The command for finding the partial fraction decomposition of a rational function *ratfunc* (in the variable x) is convert(*ratfunc*,parfrac,x). As an example, we use MAPLE V to find the integral

$$\int \frac{4x^4 + 9x^3 + 12x^2 + 9x + 4}{(x+1)(x^2+x+1)^2}\,dx.$$

```
>   rat := (4*x^4+9*x^3+12*x^2+9*x+4)
      /(x + 1)/(x^2 + x + 1)^2:
>   convert(rat,parfrac,x);
```

$$\frac{2}{x+1} + \frac{1+2x}{x^2+x+1} + \frac{1}{(x^2+x+1)^2}$$

```
>   int(",x);
```

$$2\ln(x+1) + \ln(x^2+x+1) + \frac{1}{3}\frac{2x+1}{x^2+x+1}$$
$$+ \frac{4}{9}\sqrt{3}\arctan\left(\frac{1}{3}(2x+1)\sqrt{3}\right)$$

5.8 Taylor and series expansions

The command to find the first n terms of the Taylor series expansion for $f(x)$ about the point $x = c$ is taylor(f(x),x=c,n). We compute the first five terms of the Taylor series expansion of $y = (1 - x)^{-1/2}$ about $x = 0$.

```
>   y := 1/sqrt(1-x);
```

$$y := \frac{1}{\sqrt{1 - x}}$$

```
>   taylor(y,x=0,5);
```

$$1 + \frac{1}{2}x + \frac{3}{8}x^2 + \frac{5}{16}x^3 + \frac{35}{128}x^4 + O\left(x^5\right)$$

To find a specific coefficient in a Taylor series expansion, use `coeff`.

```
>   J := product(1-x^'i','i'=1..50):
>   taylor(J^3,x=0,20);
```

$$1 - 3x + 5x^3 - 7x^6 + 9x^{11} - 11x^{15} + O(x^{20})$$

```
>   coeff(",x,15);
```

$$-11$$

To convert a *series* into a polynomial, try `convert(series, polynom)`. Also, see `?series`.

5.9   Solving differential equations

To solve the differential equation *de* involving $y = f(x)$ we use the command `dsolve(de,y)`.

```
>  f:='f':    y := f(x);
```

$$y := f(x)$$

```
>  dy := diff(y,x);
```

$$\frac{d}{dx}f(x)$$

```
>  ddy := diff(",x);
```

$$\frac{d^2}{dx^2}f(x)$$

```
>  dsolve(ddy+5*dy+6*y = sin(x)*exp(-3*x)
      ,y);
```

$$\frac{1}{2}\cos(x)e^{-3x} - \frac{1}{2}\sin(x)e^{-3x} + \_C1e^{-3x} + \_C2e^{-2x}$$

We found that the general solution to the differential equation

$$y'' + 5y' + 6y = \sin x\, e^{-3x}$$

is

$$y = \frac{1}{2}\cos(x)e^{-3x} - \frac{1}{2}\sin(x)e^{-3x} + c_1 e^{-3x} + c_2 e^{-2x},$$

where c_1 and c_2 are any constants.

Systems of differential equations can be solved in an analogous fashion. To solve the initial value problem

$$y' + z' = e^x, \qquad y(0) = 8/9$$
$$y' - 3z = x, \qquad z(0) = 10/9,$$

try

```
>   g:='g':  y:=f(x):  z:=g(x):
>   de1 := diff(y,x) + diff(z,x) = exp(x);
>   de2 := diff(y,x) - 3*z = x;
>   dsolve({de1,de2,f(0)=8/9,g(0)=10/9},
       {y,z});
```

To find series solutions, use the option type=series. Type ?dsolve to get more information and examples.

5.10   Asymptotic expansions

To find the first n terms of the asymptotic expansion of the function $f(z)$ we use the command asympt(f(z),z, n). For example, below we find the first few terms of the asymptotic expansion of the psi-function (which is the logarithmic derivative of the gamma function).

```
>   z:='z':  asympt(Psi(z),z,3);
```

$$\ln(z) - \frac{1}{2z} - \frac{1}{12z^2} + O\left(\frac{1}{z^4}\right)$$

6. GRAPHICS

MAPLE V can plot functions of one variable, planar curves, functions of two variables, and surfaces in three dimensions. It can also handle parametric plots and animations. The two main plotting functions are `plot` and `plot3d`.

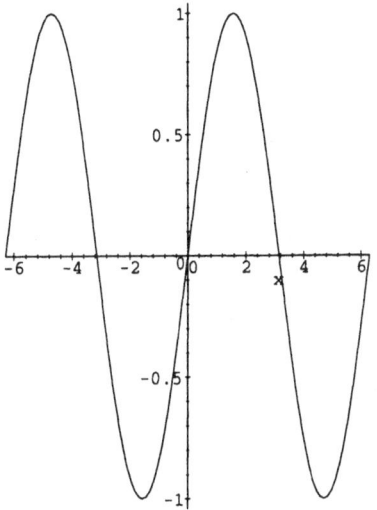

Figure 6.1 Maple plot of $y = \sin x$.

6.1   2-dimensional plotting

The syntax for plotting an expression (or function) in x is `plot(f(x),x=a..b)`. For example, to plot $\sin(x)$ for $-2\pi \le x \le 2\pi$, we type

```
>  plot(sin(x),x=-2*Pi..2*Pi);
```

The resulting plot appears in Figure 6.1.

Observe that in MAPLE V (Release 4) the plot actually appears in the current document. Now try *clicking* on the plot. A rectangular box, containing the plot, should appear. There should also be little black squares in the corners. Try holding the left mouse button down to resize the plot. Notice also that the menu bar and the context bar have changed. The Insert, Format and Options menus have been replaced by the Style, Axes, Projection, and Animation menus. The context bar has changed completely. There should be a small window containing a pair of coordinates and nine new buttons. Try clicking on each button to see its effect.

[0.5191, 0.4999]	Displays the coordinates of the point under the tracker (i.e., the point *clicked*).

 Render the plot using the usual line style.

 Render the plot using the usual point style.

 Render the plot using the polygon patch with gridlines style.

 Render the plot using the polygon patch style.

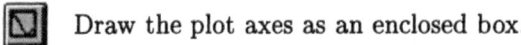	Draw the plot axes as an enclosed box.
	Draw the plot axes as an exterior frame.
	Draw the plot axes in traditional form.
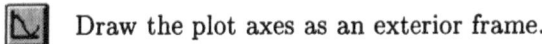	Suppress the drawing of plot axes.
	Use the same scale on both axes.

6.1.1 Restricting domain and range

Try the plot command `plot(sec(x),x=-Pi..2*Pi)`. Notice the "spikes" at $x = -\pi/2$, $\pi/2$ and $3\pi/2$ in your maple plot. These correspond to singularities of $\sec(x)$. We restrict the range to get a more reasonable plot.

```
>   plot(sec(x),x=-Pi..2*Pi,y=-5..5);
```

The resulting plot appears in Figure 6.2.

So, to plot $y = f(x)$, where $a \leq x \leq b$, and $c \leq y \leq d$, in MAPLE V we use the command `plot(f(x),x=a..b,y=c..d)`.

6.1.2 Parametric plots

To plot the curve parametrized by

$$x = f(t), \qquad y = g(t), \qquad \text{for } a \leq t \leq b,$$

we use the command `plot([f(t),g(t),t=a..b])`. The ellipse

$$x^2 + 4y^2 = 1,$$

65

can be parametrized as

$$x = \cos(t), \quad y = \frac{1}{2}\sin(t), \quad \text{where } 0 \leq t \leq 2\pi.$$

Try

```
>   plot([cos(t),1/2*sin(t),t=0..2*Pi]);
```

This should give you the desired plot.

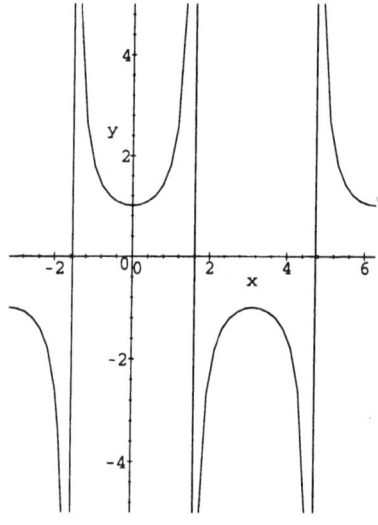

Figure 6.2 Maple plot of $y = \sec x$.

6.1.3 Multiple plots

To plot the two functions

$$y = \sqrt{x}, \qquad y = 3\log(x),$$

try

```
> plot([sqrt(x),3*log(x)],x=0..400);
```

The resulting plot is given in Figure 6.3. Each curve is plotted with a different color. Observe that our plot does not seem to illustrate the expected behaviour of the log function near $x = 0$. To get a more accurate plot, we can use the numpoints option. Try

```
> plot([sqrt(x),3*log(x)],x=0..400,
    numpoints=1000);
```

An alternative method for doing multiple plots is to use the display function in the *plots* package. Try

```
> with(plots):
> p1:=plot(sqrt(x),x=0..400):
> p2:=plot(3*log(x),x=0..400):
> display(p1,p2);
```

When defining p1 and p2, use a colon unless you want to see all the points maple uses to plot the functions. To see all the functions in the *plots* package, type

```
> with(plots);
```

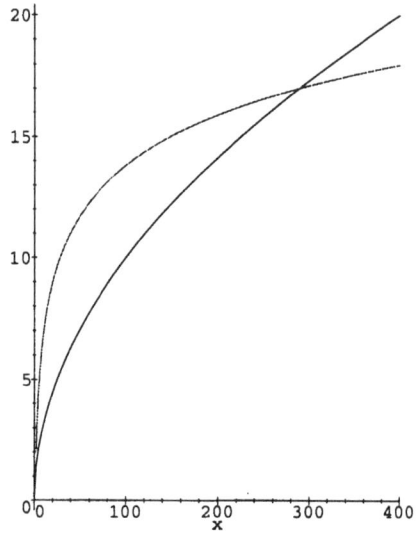

Figure 6.3 Maple plot of $y = \sqrt{x}$ and $y = 3 \log x$.

6.1.4 Polar plots

To plot polar curves we use the `polarplot` function in the *plots* package. Use the command `polarplot(f(t), t=a..b)` to plot the polar curve $r = f(\theta)$. Try

```
> with(plots):
> polarplot(cos(5*t),t=0..2*Pi);
```

68

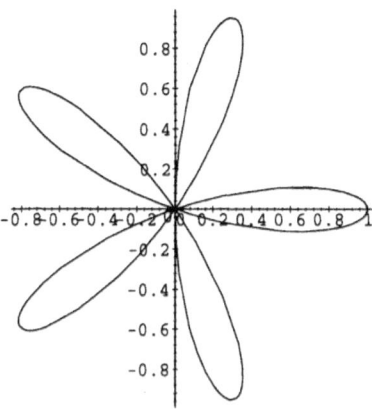

Figure 6.4 Maple plot of the polar curve $r = \cos 5\theta$.

When you try this the first time you will notice the scale on the x-axis is different to that on the y-axis. To make the scales the same, hold the first mouse button on <u>P</u>rojection and release on <u>C</u>onstrained; or, better still, click on 1∶1 .

6.1.5 Plotting implicit functions

In Section 6.1.2 we used a paramterization to plot the curve $x^2 + 4y^2 = 1$. Alternatively, we can plot implicitly defined functions using the implicitplot command in the *plots* package. Try

```
>  with(plots):
>  implicitplot(x^2+4*y^2=1,x=-1..1,
     y=-1/2..1/2);
```

69

This should agree with what we obtained before.

6.1.6 Plotting points

In MAPLE V, we plot the points

$$(x_1, y_1), (x_2, y_2), \ldots, (x_n, y_n)$$

with the command `plot([[x1,y1],[x2,y2], ..., [xn,yn]])`. Try

```
> L := [[0,0],[1,1],[2,3],[3,2],[4,-2]]:
> plot(L);
```

The resulting plot is given in Figure 6.5. Notice that MAPLE V (by default) has drawn lines between the points. To plot the points and nothing but the points, try

```
> plot(L, style=point);
```

The points correspond to plus-signs.

6.1.7 Title and text in a plot

To put a title above a plot, we use the option `title`. Try

```
> p1:=plot([sqrt(x),3*log(x)],x=0..400,
    title='The Square Root and
    log functions'):
> display(p1);
```

To add text to a plot, we use the `texplot` and `display` functions in the *plots* package. Try

```
>  p2:=textplot([[360,16,'y=3log(x)'],
      [130,10,'y=sqrt(x)']]):
>  display(p1,p2);
```

`textplot([x1,y1,string])` creates a plot with *string* positioned at (x_1, y_1).

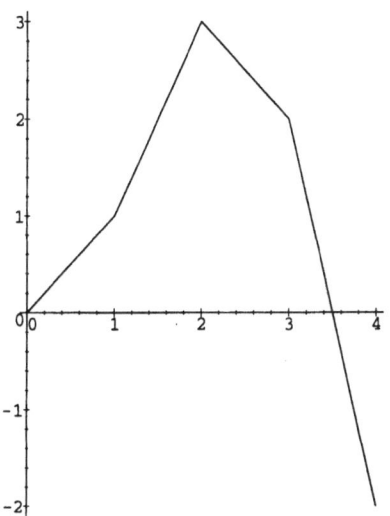

Figure 6.5 Maple plot of some data points.

6.1.8 Plotting options

The plotting options are given after the function and ranges in the **plot** command. See

71

`plot[options]` for a complete listing. Options include

axes	*frame*, *boxed*, *normal*, or *none*
discont	for plotting a discontinuous function
font	Try `font=[HELVETICA,12]`.
labelfont	font for labels on the axes
linestyle	dashed pattern for lines
numpoints	number of plotting points
resolution	horizontal display resolution
scaling	Use *constrained* for equal scale.
style	Use *point* for points.
symbol	symbol for *point* style
thickness	line thickness
title	title for the plot
titlefont	font for the title
xtickmarks	number of x-axis scale marks

6.1.9   Saving and printing a plot

There are several ways to save a plot. Any plot that is part of a worksheet will be saved when the worksheet is saved. See Sections 9.2 and 9.3. The `plotsetup` function can be used to save a plot as a file suitable for other drivers. This is done by specifying the `plotdevice` variable. Common settings for `plotdevice` are

ps	encapsulated Postscript file
jpeg	24-bit color JPEG file
hpgl	HP GL file

Here is an example.

```
>    plotsetup(ps, plotoutput='plot.ps',
        plotoptions='portrait, noborder');
>    plot(sin(x),x=-2*Pi..2*Pi);
>    interface(plotdevice=inline);
```

In this session, a plot of $y = \sin(x)$ was written to the Postscript file *plot.ps*, in portrait style with no surrounding border. The `interface` function was used so that any future plot will be within the worksheet. Otherwise, if `plotsetup` is not changed, any future plot will overwrite the file *plot.ps*.

A plot may be printed as part of the worksheet using the menu. Alternatively, it can be saved as a file and printed using a graphics driver. For example, try

```
>    plotsetup(hpgl, plotoutput='plot.hp',
        plotoptions='laserjet');
```

when printing a plot with a HP Laserjet printer. For more information, use the help commands ?plotsetup, ?plot[device].

6.2   3-dimensional plotting

The syntax for plotting an expression (or function) in two variables (say x, y) is `plot3d(f(x,y)`, `x=a..b,y=c..d)`. For example, to plot the function $z = e^{-(x^2+y^2-1)^2}$ for $-2 \leq x, y \leq 2$, we use the command

```
> plot3d(exp(-(x^2 + y^2-1)^2), x=-2..2,
    y=-2..2);
```

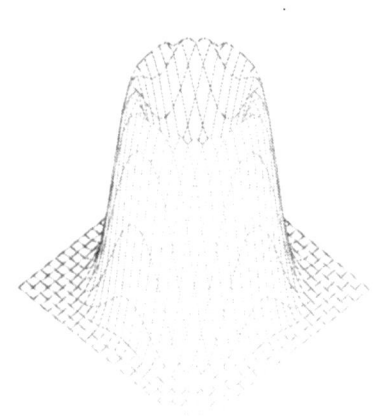

Figure 6.6 A plot of the function $z = e^{-(x^2+y^2-1)^2}$.

Observe (as before with 2-dimensional plotting) that the plot appears in the worksheet. Now try *clicking* on the plot. Notice the appearance of the Style, Colour, Axes, Projection, and Animation menus. The context bar has also changed. There should be a pair of small windows labelled ϑ and ϕ, each containing the number 45. This pair of numbers refer to a point in spherical coordinates and correspond to the orientation of the plot. There should also be thirteen new buttons. Try clicking on each button to see its effect.

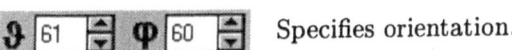 Specifies orientation.

Render the plot using the polygon patch style with gridlines.

Render the plot using the polygon patch style.

Render the plot using the polygon patch and contour style.

Render the plot using the hidden line removal style.

Render the plot using the contour style.

Render the plot using the wireframe style.

Render the plot using the point style.

Draw the plot axes as an enclosed box.

Draw the plot axes as an exterior frame.

Draw the plot axes in traditional form.

Suppress the drawing of plot axes.

Use the same scale on each axis.

Redraw the plot.

Now, hold the first mouse button down on the plot. A cube should appear. Drag the mouse so that the cube rotates to the desired position. Notice that the value of (ϑ, ϕ) has changed. Double click on the cube or click on **R** to redraw the plot. Below is a plot obtained by clicking on 🌐 and 🔲 and selecting $(\vartheta, \phi) = (22, 67)$.

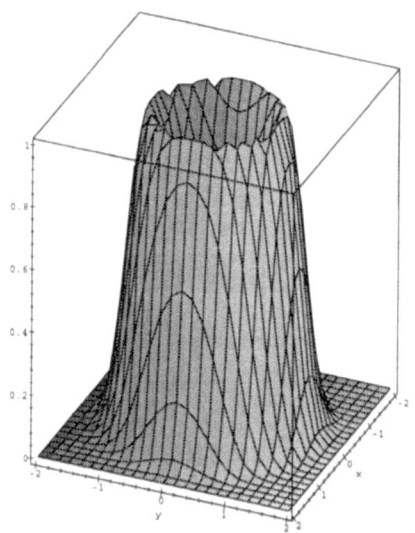

Figure 6.7 A Maple plot with boxed axes.

Now, try clicking 🌐 to see some hidden detail of the plot. You might use the **grid** option to increase the number of contours plotted. Try

```
> plot3d(exp(-(x^2 + y^2)^2), x=-2..2,
    y=-2..2,grid=[50,50]);
```

Don't forget to either double click or click on .

6.2.1   Parametric plots

To plot the surface parametrized by

$$x = f(u,v), \quad y = g(u,v), \quad z = h(u,v),$$

where $a \leq u \leq b$, $c \leq v \leq d$; use the command
plot3d([f(u,v), g(u,v), h(u,v)], u=a..b, v=c
..d). For example, the hyperboloid

$$x^2 + y^2 - z^2 = 1,$$

may be parametrized by

$$x = \sqrt{1 + u^2} \cos t, \quad y = \sqrt{1 + u^2} \sin t, \quad z = u,$$

where $-\infty < u < \infty$ and $0 \leq t \leq 2\pi$. Try

```
>  plot3d([sqrt(1+u^2)*cos(t),sqrt(1+u^2)
     *sin(t),u], u=-1..1, t=0..2*Pi);
```

A plot with $(\vartheta, \phi) = (45, 60)$ is given in Figure 6.8.

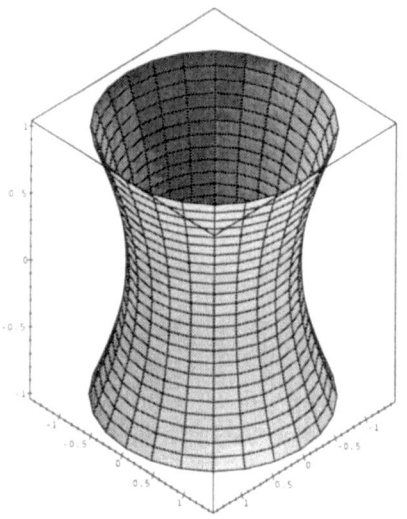

Figure 6.8 Maple plot of an hyperboloid.

6.2.2 Multiple plots

To plot the two functions

$$z = e^{-x^2 - y^2},$$
$$z = x + y + 1,$$

try

```
>   plot3d({exp(-x^2-y^2),x+y+1},x=-2..2,
      y=-1..1);
```

with $(\vartheta, \phi) = (120, 45)$. As with 2-dimensional
plotting, multiple 3-dimensional plots can be pro-
duced using the **display** function in the *plots* package. Try

```
>   with(plots):
>   p1:=plot3d(exp(-x^2-y^2),x=-2..2,
      y=-1..1):
>   p2:=plot3d(x+y+1,x=-2..2,y=-1..1):
>   display(p1,p2);
```

6.2.3   Space curves

To plot the space curve

$$x = f(t), \quad y = g(t), \quad z = h(t),$$

where $a \leq t \leq b$, we use the **spacecurve**
function in the *plots* package. The command
is **spacecurve([f(t),g(t),h(t)],t=a..b)**. We
plot the helix

$$x = \cos t, \quad y = \sin t, \quad z = t.$$

Try

```
>   with(plots):
>   spacecurve([cos(t),sin(t),t],t=0..4*Pi,
      numpoints=200);
```

79

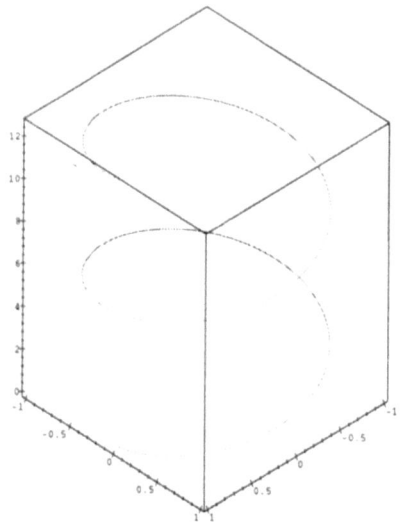

Figure 6.9 Maple plot of a helix.

6.2.4 Contour plots

The graph of a function of two variables may be visualized with a 2-dimensional contour plot. To produce contour plots, we use the functions `contourplot` and `contourplot3d` in the *plots* package. `Contourplot3d` "paints" the contour plot on the corresponding surface. Try

```
> with(plots):
```

```
>   contourplot(exp(-(x^2+y^2-1)^2),
      x=-(1.3)..(1.3), y=-(1.3)..(1.3),
      filled=true, coloring=[blue,red]);
>   contourplot3d(exp(-(x^2+y^2-1)^2),
      x=-(1.3)..(1.3), y=-(1.3)..(1.3),
      filled=true, coloring=[blue,red]);
```

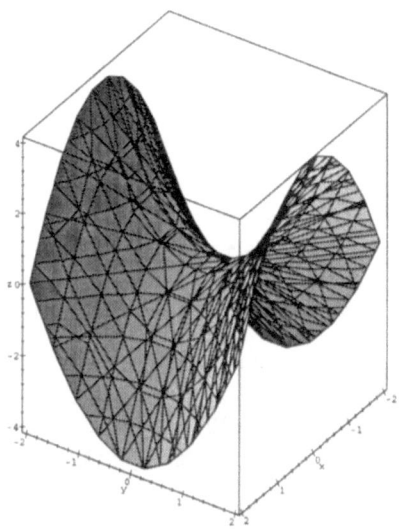

Figure 6.10 Maple plot of a hyperbolic paraboloid.

6.2.5 Plotting surfaces defined implicitly

To plot the surface defined implicitly by the equation

$$f(x, y, z) = c,$$

81

use the command `implictplot3d(f(x,y,z)=c,` `x=a..b,y=d..e,z=g..h)` in the *plots* package. For example, to plot the hyperbolic paraboloid

$$y^2 - x^2 = z,$$

try

```
> with(plots):
> implicitplot3d(y^2 - x^2 = z, x=-2..2,
    y=-2..2, z=-4..4);
```

The resulting plot is given in Figure 6.10.

In Section 6.2.1 we obtained a plot of the surface

$$x^2 + y^2 - z^2 = 1,$$

by using a parametrization. This time, try

```
> implicitplot3d(x^2 + y^2 - z^2 = 1,
    x=-1..1, y=-1..1, z=-1..1);
> implicitplot3d(x^2 + y^2 - z^2 = 1,
    x=-2..2, y=-2..2, z=-1..1);
```

Notice how care must be taken in choosing the range for each variable.

6.2.6 Title and text in a plot

A title or text may be inserted in a 3-dimensional plot in the same way it was done in Section 6.1.7 for 2-dimensional plots. Try

```
>  with(plots):
>  p1:=plot3d(exp(-(x^2+y^2-1)^2),x=-2..
2,y=-2..2, font=[TIMES,ROMAN,12],titlefont
=[HELVETICA,BOLD,10], title='The surface
z=exp(-(x^2+y^2-1)^2)'):
>  p2:=textplot3d([0,1.1,1,'Circular
     Rim'], align=RIGHT,color=BLUE):

>  display(p1,p2);
```

6.2.7 3-dimensional plotting options

The options **axes**, **font**, **labels**, **labelfont**, **linestyle**, **numpoints**, **scaling**, **symbol**, **thickness**, **title**, **titlefont**, and **view** should work like they did for 2-dimensional plotting (see Section 6.1.8). Other options are **ambientlight**, **color**, **contours**, **coords**, **gridstyle**, **light**, **lightmodel**, **orientation**, **projection**, **shading**, and **style**. See **?plot3d[options]** for more information.

6.3 Animation

MAPLE V is capable of animating 2- and 3-dimensional plots. The two animation functions are **animate** and **animate3d**. These are in the *plots* package. For fixed t we consider the function

$$f_t(x) = \frac{1}{1+xt}.$$

We may examine the behaviour of this function as t changes using **animate**. Try,

```
>   with(plots):
>   animate(1/(1+x*t),x=0..10,t=0..1,
        frames=10);
```

A plot of $f_0(x) = 1$ should appear in the work-sheet. Now click on the plot. A new context bar should appear containing a window for coordinates and nine new buttons similiar to those on a cassette tape player. Try clicking on each button to see its effect.

■	Stop the animation.
▶	Play the animation.
⇥	Move to the next frame.
←	Set the animation direction to be backward.
→	Set the animation direction to be forward.
◀◀	Decrease the speed of the animation.
▶▶	Increase the speed of the animation.
⇉	Set animation to run in single-cycle mode.
↻	Set animation to run in continuous-cycle mode.

Now click on ▶ to play the animation. The **frames** option allows you to set the number of separate frames in the animation. To view each frame, click on ⏭. Try setting frame=50. Now try

```
>  animate([Pi/2*sin(t*(u+1)),sin(2*t)
   *sin(Pi/2*sin(t*u+t)),t=-2*Pi..2*Pi],
   u=0..1,frames=20,numpoints=200,
   color=blue);
```

The 3-dimensional animation command is animate3d. The surface

$$x^2 - y^2 = z,$$

may be parametrized by

$$x = r\cos t, \quad y = r\sin t, \quad z = r^2\cos 2t.$$

Try animating a rotation of this surface

```
>  with(plots):
>  animate3d([r*cos(t+a),r*sin(t+a),r^2
   *cos(2*t)],r=0..1,t=0..2*Pi, a=0..3,
   frames=10,style=patch,title='The
   Rotating Saddle');
```

A little adjusting creates a *Flying Pizza*

```
>  animate3d([r*cos(t+a),r*sin(t+a),r^2
   *cos(2*t)+sin(a)],r=0..1,t=0..2*Pi,
   a=0..2*Pi,frames=10,style=patch,
   title='The Flying Pizza');
```

85

Try clicking on ⟳ to set your pizza in continuous motion.

7. LINEAR ALGEBRA

MAPLE V can do symbolic and floating point matrix computations. The linear algebra functions are contained in the *linalg* package. Try

```
>   ?linalg
```

to see a list of these functions.

7.1 Vectors, arrays, and matrices

Matrices and vectors are data types defined within the *linalg* package. It is necessary to load the *linalg* package before creating matrices and vectors. In MAPLE V a matrix is a two-dimensional array. It would be a good idea to reread Section 4.5 on arrays. We give some examples on creating vectors and matrices.

```
>   with(linalg):
>   v:=vector([1,2,3]);
```

$$v := [1, 2, 3]$$

```
>    A:=matrix(2,3,[a,b,c,d,e,f]);
```

$$\begin{bmatrix} a & b & c \\ d & e & f \end{bmatrix}$$

```
>  v;
```

$$v$$

```
>  A;
```

$$A$$

```
>  print(v);
```

$$[1, 2, 3]$$

```
>  print(A);
```

$$\begin{bmatrix} a & b & c \\ d & e & f \end{bmatrix}$$

We used the **vector** and **matrix** functions in the
linalg package to define the 3-dimensional vector
v and the 2×3 matrix A. Notice that typing
v or **A** did not cause the vector or matrix to be
displayed. We displayed them using the **print**
command. Also, try

```
>  op(A);
>  eval(A);
```

It is possible to enter a matrix interactively using
the **entermatrix** command.

```
>  with(linalg):
>  B := matrix(2,2);
```

$$B := \mathbf{array}(1\,..\,2, 1\,..\,2, [\,])$$

```
>  entermatrix(B);
enter element 1,1 >   12;
enter element 1,2 >   13;
enter element 2,1 >   14;
enter element 2,2 >   15;
```

$$\begin{bmatrix} 12 & 13 \\ 14 & 15 \end{bmatrix}$$

Note that the semi-colon must still be used when entering matrix elements.

A fun way to create matrices is to use a function $f(x,y)$ of two variables. The function matrix(m,n,f) produces the $m \times n$ matrix whose (i,j)-th entry is $f(i,j)$. Try

```
>  f := (i,j) -> x^(i*j);
>  A := matrix(4,4,f);
>  factor(det(A));
```

7.2   Matrix operations

MAPLE V can do the usual matrix operations of addition, multiplication, scalar multiplication, inverse, transpose, and trace.

Matrix operation	Mathematical notation	MAPLE V notation
Addition	$A + B$	A + B
Subtraction	$A - B$	A - B
Scalar multiplication	$c\,A$	c*A
Matrix multiplication	$A\,B$	A &* B or multiply(A,B)
Matrix power	A^n	A^n
Inverse	A^{-1}	A^(-1) or 1/A or inverse(A)
Transpose	A^T	transpose(A)
Trace	$\operatorname{tr} A$	trace(A)

Look at the following example:

```
> with(linalg):
> A:=matrix(2,2,[1,2,3,4]):
> B:=matrix(2,2,[-2,3,-5,1]):
> A+B;
```

$$A + B$$

```
> evalm(");
```

$$\begin{bmatrix} -1 & 5 \\ -2 & 5 \end{bmatrix}$$

Notice that we had to use the function **evalm** to display the matrix $A + B$. Now try the following:

```
>    with(linalg):
>    A:=matrix(2,3,[1,2,3,4,5,6]);
>    B:=matrix(3,2,[2,4,-7,3,5,1]);
>    C:=matrix(2,2,[1,-2,-3,4]);
>    A&*B;
>    evalm(");
>    multiply(A,B);
>    evalm(A&*B-2*C);
```

Now check your results with pencil and paper. You should have found that

$$A B - 2 C = \begin{bmatrix} 1 & 17 \\ 9 & 29 \end{bmatrix}$$

7.3   Elementary row operations

MAPLE V can perform all the elementary row and column operations.

Elementary row operation	Operational notation	MAPLE V notation
Swap two rows	$R_i \longleftrightarrow R_j$	swaprow(A,i,j)
Multiply a row by constant	$R_i \longrightarrow c R_i$	mulrow(A,i,c)
Add a multiple of one row to another	$R_j \longrightarrow c R_i + R_j$	addrow(A,i,j,c)

Let

$$A = \begin{bmatrix} 1 & 1 & 3 & -3 \\ 5 & 5 & 13 & -7 \\ 3 & 1 & 7 & -11 \end{bmatrix}$$

Try the following elementary row operations to reduce A to *row echelon form*.

```
>   with(linalg):
>   A:=matrix(3,4,[1,1,3,-3,5,5,13,-7,3,1,
    7,-11]);
>   A1:=addrow(A,1,2,-5);
>   A2:=addrow(A1,1,3,-3);
>   A3:=mulrow(A2,3,-1/2);
>   A4:=swaprow(A3,2,3);
>   A5:=mulrow(A4,3,-1/2);
```

The last matrix should be

$$A5 = \begin{bmatrix} 1 & 1 & 3 & -3 \\ 0 & 1 & 1 & 1 \\ 0 & 0 & 1 & -4 \end{bmatrix},$$

which is in row echelon form. In this next section we will see how to check this result using Gaussian elimination.

In MAPLE V the elementary columns operations are done in a similar fashion. This time the functions are **swapcol**, **mulcol**, and **addcol**.

7.4   Gaussian elimination

MAPLE V can do Gaussian and Gauss-Jordan elimination. The relevant functions are gausselim and gaussjord. In the previous section we reduced a matrix to echelon form using elementary row operations. Check our result using gausselim and gaussjord.

```
>   with(linalg):
>   A:=matrix(3,4,[1,1,3,-3,5,5,13,-7,3,1,
      7,-11]);
>   gausselim(A);
>   gaussjord(A);
```

7.5   Inverses and determinants

To find the inverse of a matrix and its determinant, we use the functions inverse and det.

```
>   with(linalg):
>   A:=matrix(3,3,[1,1,3,5,5,13,3,1,7]):
>   det(A);
```

$$-4$$

```
>   B:=inverse(A);
```

$$B := \begin{bmatrix} \frac{-11}{2} & 1 & \frac{1}{2} \\ -1 & \frac{1}{2} & \frac{-1}{2} \\ \frac{5}{2} & \frac{-1}{2} & 0 \end{bmatrix}$$

We first found that $\det(A) = -4 \neq 0$ so that A is invertible; then found that

$$A^{-1} = \begin{bmatrix} \frac{-11}{2} & 1 & \frac{1}{2} \\ -1 & \frac{1}{2} & \frac{-1}{2} \\ \frac{5}{2} & \frac{-1}{2} & 0 \end{bmatrix}.$$

Now check your answer:

```
>   evalm(B&*A);
```

7.6   Row space, column space, nullspace

Let

$$A = \begin{bmatrix} 1 & 4 & -10 & 3 & -3 \\ 10 & 41 & -102 & 30 & -31 \\ -9 & -19 & 56 & -27 & 10 \end{bmatrix}.$$

We can use MAPLE V to find the rank of A and to find bases for the row space, column space, and null space. The relevant MAPLE V functions are rank, rowspace, colspace, and nullspace.

```
>   with(linalg):
>   A:=matrix(3,5,[1,4,-10,3,-3,10,41,-102,
      30,-31,-9,-19,56,-27,10]):
>   rank(A);
```
                                2

```
> rowspace(A);
```

$$\{[1, 0, -2, 3, 1], [0, 1, -2, 0, -1]\}$$

```
> colspace(A);
```

$$\{[1, 0, -179], [0, 1, 17]\}$$

```
> nullspace(A);
```

$$\{[-1, 1, 0, 0, 1], [-3, 0, 0, 1, 0], [2, 2, 1, 0, 0]\}$$

7.7 Eigenvectors and diagonalization

Let

$$A = \begin{bmatrix} 177 & 77 & -28 \\ -546 & -236 & 84 \\ -364 & -154 & 51 \end{bmatrix}$$

We use **eigenvals** to find the eigenvalues of A.

```
> with(linalg):
> A:=matrix(3,3,[177,77,-28,-546,-236,
    84,-364,-154,51]):
> eigenvals(A);
```

$$2, \quad -5, \quad -5$$

We see that A has two eigenvalues $\lambda = 2$ and $\lambda = -5$ (multiplicity 2). Now, let's find a basis for each eigenspace using `eigenvects`.

```
>  eigenvects(A);
```

$$[2, 1, \{[1, -3, -2]\}],$$
$$[-5, 2, \{[1, 0, 13/2], [0, 1, 11/4]\}]$$

We see that the eigenspace corresponding to $\lambda = 2$ is one dimensional and that $\{[1, -3, -2]\}$ is a basis. For $\lambda = -5$, the eigenspace is two dimensional and a basis is $\{[1, 0, 13/2], [0, 1, 11/4]\}$. Hence, we have found three independent eigenvectors and A is diagonalizable. So, we let

$$P = \begin{bmatrix} 1 & 2 & 0 \\ -3 & 0 & 4 \\ -2 & 13 & 11 \end{bmatrix}$$

Then $P^{-1}AP$ should be a diagonal matrix. Try

```
>  P:=matrix(3,3,[1,2,0,-3,0,4,-2,13,11]);
>  evalm(inverse(P)&*A&*P);
```

Did you get a diagonal matrix? Alternatively, we can use `jordan` to diagonalize A. Try

```
>  jordan(A, 'P');
>  print(P);
```

This time you should get the same diagonal matrix but the matrix P is different (since it is not unique).

MAPLE V can also compute eigenvalues and eigenvectors for complex matrices and matrices with floating point entries. Try

```
> A:=matrix(2,2,[1.0,2.0,3.0,4.0]);
> eigenvals(A);
> eigenvects(A);
> B:=matrix(2,2,[1+10*I,-8*I,12*I,
    1-10*I]);
> eigenvals(B);
> eigenvects(B);
> jordan(B, 'P');
> print(P);
```

7.8   Jordan form

We used the function **jordan** in the previous section. In general, **jordan** gives the Jordan canonical form of a square matrix. Try

```
> C:=matrix(4,4,[10,10,-14,15,0,3,0,0,8,
    1,-13,8,1,-8,-2,-4]);
> jordan(C,'Q');
> evalm(1/Q&*C&*Q);
```

7.9   Random matrices

The MAPLE V function **randmatrix(m,n)** produces a random integer $m \times n$ matrix with entries between -99 and 99. Try

```
>   with(linalg):
>   A:=randmatrix(3,3);
>   B:=randmatrix(3,3,unimodular);
>   C:=randmatrix(3,3,unimodular);
>   F:=evalm(transpose(C)&*B);
>   inverse(F);
```

Can you see why the matrix F^{-1} must have integer
entries?

7.10   More *linalg* functions

augment	augmented matrix
backsub	back substitution
blockmatrix	block matrix
charmat	characteristic matrix
cond	standard condition number
copyinto	copies a matrix into another
crossprod	crossproduct of two vectors
curl	curl of a vector field
diag	block diagonal matrix
diverge	divergence of a vector field
dotprod	dot-product of two vectors
geneqns	generate system of equations
genmatrix	generate augmented matrix
grad	gradient of a function
GramSchmidt	Gram-Schmidt orthog. process
innerprod	innerproduct $\mathbf{u}^T A \mathbf{v}$
jacobian	Jacobian matrix
JordanBlock	Jordan block matrix
leastsqrs	least squares problem

linsolve	solve a linear system
LUdecomp	*LU*-decomposition
matadd	computes a matrix sum
minpoly	minimal polynomial of a matrix
pivot	pivot a matrix
potential	potential function
QRdecomp	*QR*-decomposition of a matrix
rowdim	number of rows
singularvals	singular values of a matrix
stack	stacks two matrices
submatrix	extract a submatrix
vecpotent	vector potential of a vector field
wronskian	Wronskian matrix

8.   MAPLE V PROGRAMMING

MAPLE V is a programming language as well as an interactive symbolic calculator. It is possible to solely use MAPLE V interactively and not bother with its programming features. However, it is well worth the effort in developing some programming skills. The MAPLE V language is much is easier to learn than the traditional programming languages and you do not need to be an expert programmer to master it. You will appreciate the real power of MAPLE V when you learn some of the basic MAPLE V language and use it in combination with its interactive features. If you have gotten this far into the book, you are already familiar with many MAPLE V commands and the step to MAPLE V programming is not a big one.

8.1 Conditional statements

A <u>conditional statement</u> has the form

```
if     condition    then
       statseq
else
       statseq
fi:
```

Here *statseq* is a sequence of statements separated by semi-colons (or colons). For example,

```
>    x:=1;
```
$$x := 1$$

```
>    if x>0 then
>       y:=x+1
>    else
>       y:=x-1
>    fi:
>    y;
```
$$2$$

This conditional statement means that if $x > 0$ then $y = x + 1$, but if $x \leq 0$ then $y = x - 1$. In the session $x = 1 > 0$ so $y = x + 1 = 2$.

8.2 Loops

A <u>loop statement</u> has the form

```
for    var    from    num1    to num2    do
       statseq
```

```
od:
```

For instance, we can print out the numbers from
1 to 10.

```
>   for i from 1 to 10 do
>     print(i);
>   od:
```

We can also sum the integers from 1 to 10 the
old-fashioned way.

```
>   x:=0:
>   for i from 1 to 10 do
>     x:=x+i:
>   od:
>   x;
```

$$55$$

Hence the sum is 55. We can check our answer.

```
>   1+2+3+4+5+6+7+8+9+10;
```

$$55$$

```
>   sum('i','i'=1..10);
```

$$55$$

We can change the step-size in a loop by using **by**.

```
>   for i from 2 by 3 to 20 do
>       print(i);
>   od;
```

8.3 Procedures

A MAPLE V <u>procedure</u> has the form

```
proc(nameseq)
  local   nameseq;
  global  nameseq;
      statseq;
end:
```

The `local` and `global` statements are optional.
See the next section. Here is an example.

```
>   f:=proc(x)
>      local z;
>      if x>=0 and x<=1 then
>        z:=x^2:
>      else
>        z:=1-x:
>      fi:
>      RETURN(z);
>   end;
```

```
f := proc (x) local z; if 0 <= x and x <= 1
then z := x^2 else :z := 1-x fi; RETURN(z) end
```

The input of the procedure (or function) f is a
number x. Using a conditional statement, we were
able to define the function

$$f(x) = \begin{cases} x^2 & \text{if } 0 \le x \le 1, \\ 1 - x & \text{otherwise.} \end{cases}$$

We check that the function works.

```
>    f(1/2);
```
$$\frac{1}{4}$$
```
>    f(2);
```
$$-1$$

Now try plotting the function.

```
> plot(f,-2..2);
```

Remember this syntax. When f is a *proc*, the command `plot(f(x),x=-2..2)` will not work.

We now examine a more complicated example. The following procedure $\text{trap}(f,a,b,n)$ computes an approximation of the definite integral $\int_a^b f(x)\,dx$ using the trapezoidal rule with n divisions. Type it in.

```
>    trap:=proc(f,a,b,n)
>    local s,i,ds,exact,x ;
>    s:=0:
>    for i from 1 to n-1 do
>       s:=s+f(a+i*(b-a)/n):
>    od:
>    s:=2*s + f(a) + f(b):
>    ds:=s*(b-a)/2/n:
>    exact:=int(f(x),x=a..b):
>    print('The integral by the
    trapezoidal rule with n=',n,' is ');
>    print(evalf(ds));
```

102

```
>     print('The exact integral is ',
      exact,' = ',evalf(exact));
>     print('Error=',evalf(abs(ds-exact)));
>     RETURN(evalf(ds)):
>   end:
```

Adding **print** statements is very handy for debugging a progam. To compute the integral $\int_1^2 \frac{1}{x}\, dx$ using the trapezoidal rule, try

```
>     f:=x->1/x;
>     trap(f,1,2,10);
>     trap(f,1,2,100);
```

8.4   Local and global variables

If the **local** statement is not used in a MAPLE V *proc*, then all variables within the *proc* are declared *local* by default. To change the default we use the **local** and **global** statements.

```
>     g := proc(x,y)
>       local z,i;
>       global v,w;
>       if x*y>1 then
>         v:=x+y:
>       else
>         w:=x-y:
>       fi:
>       RETURN(x*y);
>     end;
```

103

Now try some examples to see what this *proc* is doing.

```
>    g(2,3);
>    v,w;
>    g(1/2,1/3);
>    v,w;
```

Do you see what's going on? Each time **g** is called, the global value of *v* or *w* is changed depending on the input (x, y).

8.5   Reading and saving *procs*

Although the *editing* features of MAPLE V are getting better and better with each release, it is usually more convenient and wiser to write MAPLE V programs using an editor and save them in ordinary text files. For instance, instead of typing the proc **trap** (given in Section 8.3) directly into a worksheet within maple, it would be better to create it using an editor in, say, the file *trap*. The MAPLE V **read** function is used to read a file into a maple session. We give an example for *Windows*. If this file was in the sub-directory *myprogs* within the *maplev4* directory, try

```
>    read 'c:\\maplev4\\myprogs\\trap';
```

and then **trap** is ready for use. A variant of this should work on other platforms. For instance, in the *unix* version try

```
>    read trap;
```

if your MAPLE V session was started in the same
directory.

8.6 Viewing built-in MAPLE V code

One of the great features of MAPLE V is that
most of the built-in functions are written in the
MAPLE V programming language and the code is
accessible to the user. To see how MAPLE V de-
fines the Gamma function, try

```
>    interface(verboseproc=2);
```

```
>    op(GAMMA);
```

9. SAVING AND READING FILES

In Section 8.5 we saw that the MAPLE V read
command may be used to read in programs in a
MAPLE V session. In this chapter we examine the
ways the following may be saved and read: (1)
variables, (2) sessions, and (3) worksheets. Also
we will examine the different ways in which MAPLE
V worksheets may be exported.

9.1 Saving a Maple session

A MAPLE V session may saved through the
File menu by releasing on Save As ... or by
clicking on ▣. The options are then Maple
Worksheet, Maple Text, Text, and LaTeX Source.

The default is **Maple Worksheet**. The file extension for a maple worksheet is *mws*. If you saved your session as *first.mws* then, in a later session, you may open this worksheet by selecting <u>O</u>pen ... or by clicking on 📂. When this worksheet is open, the whole worksheet is visible but the values of variables have not been assigned. The values of variables may saved using the **save** command.

```
>   x:=5;
>   y:=7;
>   z:=int(1/u,u);
>   save 'first.m';
>   save x,y,part1;
```

In the session above, all the variables were saved in the maple binary file *first.m*. The values of x and y were saved in the text file *part1*.

See ?open, ?close, ?appendto, ?writeto, and ?writedata for other methods of writing to files.

9.2   Reading MAPLE V **programs**

See Section 8.5 on reading MAPLE V procs. MAPLE V programs may be read in the same manner. An existing MAPLE V worksheet may opened under the <u>F</u>ile menu by selecting <u>O</u>pen ... or by clicking on 📂.

Text files and *.m* files may be read with the **read** command. We read two files created in the last section:

106

```
>  read 'first.m';
>  read part1;
```

When *first.m* is read, the values of all the variables
x, y, and z are assigned but not displayed. When
the text file **part1** is read, the variables x and y
are assigned their previous values and displayed.

See **?readdata**, **?readline**, and **?sscanf** for
reading data.

9.3   Saving worksheets and LaTeX

In Section 9.1 we saw how a maple worksheet
may be saved as a *.mws* file and opened in a later
session. A worksheet may also be saved as a plain
text or LaTeX file. In the File menu, select Export
As ... and then select either Plain Text ... , Maple
Text ... , or LaTeX .... To convert maple output
into LaTeX, use the **latex** function. Try

```
>  with(linalg):
>  A:=matrix(3,3,(i,j)->sin(Pi*i*j/6));
>  latex(A);
```

10.   DOCUMENT PREPARATION

MAPLE V (Release 4) has many new features
for creating documents. It is now possible to add
maple output to text and create technical documents. There are also facilities for adding head-
ings, changing fonts, inserting expandable subsec-
tions, bookmarks, and hyperlinks.

We now demonstrate some of these features with a specific example. Suppose we have the following

Problem. *Reduce the weight of a ball-bearing with diameter 2 cm by 50% by drilling a hole through the center. Determine the diameter of the required drill-bit.*

This problem can be solved easily in MAPLE V by computing a certain integral and solving an equation. Start maple and type in the following.

```
>   v:=Int(4*Pi*x*sqrt(1-x^2),x=0..r);
```

$$v := \int_0^r 4\pi x \sqrt{1 - x^2} \, dx$$

```
>   v:=value(v);
```

$$-\frac{4}{3} \left(1 - r^2\right)^{3/2} \pi + \frac{4}{3} \pi$$

```
>   rrs:=solve(v=2*Pi/3,r);
```

$$rrs := \text{RootOf}(-12\_Z^4 + 12\_Z^2 - 3 + 4\_Z^6,$$
$$-0.6083087005), \quad \text{RootOf}(-12\_Z^4 + 12\_Z^2 - 3$$
$$+ 4\_Z^6, \quad 0.6083087005)$$

```
>   convert(rrs[2],radical);
```

$$\frac{1}{2} \sqrt{4 - 2\, 2^{1/3}}$$

```
>  radimp("*2);
```

$$\sqrt{2}\,\sqrt{2 - 2^{1/3}}$$

```
>  evalf(");
```

$$1.2116617400$$

The desired diameter is

$$2r = \sqrt{2}\,\sqrt{2 - 2^{1/3}} \approx 1.212\,\text{cm}.$$

You may be wondering what is going on in this problem. We can make a much clearer document by adding text.

10.1   Adding text

First we add some text to our document. Click the cursor on the first line of maple input. Then in the Insert menu, select Execution Group and Before Cursor. A maple prompt > should appear above the first line of input. Now click on $\boxed{\text{T}}$ and type

```
Reduce the volume of a ball bearing
with diameter 2 cm by 50% by
drilling a hole through the center.
Determine the diameter of the
required drill-bit.
```

To create a new paragraph, click on ▶ and then
T. Now type

```
First we observe that the ball bearing
is the solid obtained by rotating a
circle of radius 1cm about the y-axis.
If we let r be the radius of the drill-
bit then, by the shell method, the
volume of material removed is given by
```

Now we would like to add some in-line math.

10.2   Inserting math into text

In the Insert menu, select Math Input and a
red ? should appear. Type

```
Int(4*Pi*x*sqrt(1-x^2),x=0..r)
```

What was maple input should now appear as math
in your document. Click to the right of the math
and click on T and type

```
We compute the integral
```

Let's add a title.

10.3   Adding titles and headings

Click on the first line of the worksheet. In
the Insert menu, select Execution Group and
Before Cursor. Then click on T. In the box
P Normal ▼ select Title. Now type

The Ball Bearing Problem

The document should now have a title. Press enter and type your name

William E. Wilson

Your name should now be underneath the title. Press enter again. To make a heading this time, we select **Heading 2**. Type

Statement of the problem

To underline this heading, click on **u**. Now make a heading entitled **Solution** for the next paragraph.

Let's move some of the maple computations into a new subsection.

10.4 Creating a subsection

Use the first mouse button to highlight the maple inputs

```
v:=Int(4*Pi*x*sqrt(1-x^2),x=0..r);
```

and

```
v:=value(v);
```

together with their output. Now click on ➡. A little button ⊟ should appear. Try clicking on it. Pretty neat! Now see if you can add a heading to this subsection using the **Heading 3** selection.

Now we shall add some more text and math by *cutting and pasting*.

10.5 Cutting and pasting

First we create a new region. Click on the vertical bar attached to ⊟ and click on ▶ and then ⊤. There should now be a new text region below the new subsection. Now type

```
Our computation gave
```

At this point, we would like to add an equation to our document. This time we will use the mouse to *cut and paste*. First click on ▶ and type

```
> 'v' =
```

Now, instead of retyping maple input, we move the cursor to the maple output above and use the mouse to highlight

$$\boxed{-\frac{4}{3}\left(1-r^2\right)^{3/2}\pi + \frac{4}{3}\pi}$$

Use the mouse or hot-keys to copy the selection and paste it to the right of the equal-sign. The hot-keys are system dependent. In Windows, use *control-c* to copy and *control-v* to paste. Observe how the displayed math has been converted to maple input. Now type a semi-colon and press enter:

```
>    'v' =-4/3*(1-r^2)^(3/2)*Pi+4/3*Pi;
```

$$v = -\frac{4}{3}\left(1 - r^2\right)^{3/2}\pi + \frac{4}{3}\pi$$

Now use the mouse to highlight the maple input line

```
>    'v' =-4/3*(1-r^2)^(3/2)*Pi+4/3*Pi;
```

and hit *control-x* (or *delete*) and this line should now be erased. Finally, add enough text and equations so that the document is complete. A rendition of how it might appear is given below.

The Ball Bearing Problem

William E. Wilson

Statement of the problem

Reduce the volume of a ball bearing with diameter 2 cm by 50% by drilling a hole through the center. Determine the diameter of the required drill-bit.

Solution

First we observe that the ball bearing is the solid obtained by rotating a circle of radius 1cm about the y-axis. If we let r be the radius of the drill-bit then, by the shell method, the volume v of material removed is given by $\int_0^r 4\pi x\sqrt{1-x^2}\,dx$.

We compute the integral.

▣ *Computation*

> `v:=Int(4*Pi*x*sqrt(1-x^2),x=0..r);`

$$v := \int_0^r 4\pi x\sqrt{1-x^2}\,dx$$

> `v:=value(v);`

$$v := -\frac{4}{3}\left(1-r^2\right)^{3/2}\pi + \frac{4}{3}\pi$$

Our computation gave

$$v = -\frac{4}{3}\left(1-r^2\right)^{3/2}\pi + \frac{4}{3}\pi$$

We solve the equation

$$-\frac{4}{3}\left(1-r^2\right)^{3/2}\pi + \frac{4}{3}\pi = \frac{2}{3}\pi$$

⊞ *Computation*

to find that the required diameter is

$$2r = \sqrt{2}\sqrt{2 - 2^{1/3}}$$

which is approximately 1.212 cm.

10.6 Bookmarks and hypertext

A *bookmark* is a name that marks a location
in a worksheet. Selecting this name will move the
cursor to the specified location. To create a bookmark at the last equation in our document, click
the cursor on the equation. Then, in the View
menu, select Bookmarks and then Edit Bookmark
. . . . An *Add or Modify Bookmark* window should
appear. In the Bookmark Text box, type a word,
say, ANSWER and click on OK. Although the worksheet appears no different, it now has a single
bookmark. We may access this bookmark by selecting Bookmarks in the View menu. Now ANSWER
should appear in the submenu. Select ANSWER and
the cursor will move to the specified location. Try
moving the cursor to a different place in the worksheet and select ANSWER again.

Now we will use our bookmark to create a hyperlink in our worksheet. A *hyperlink* is a link from
one location in the worksheet to a different location in the worksheet or to a different worksheet
altogether. The presence of a hyperlink is indicated by green underlined text. Clicking on this
text will move the cursor to the new location. In
our worksheet we will attach a hyperlink from the
word *diameter* in the statement of the problem to
our bookmark ANSWER.

Move the cursor to the word diameter near the
top of the worksheet and in the Insert menu se-

lect <u>H</u>yperLink ... . A *HyperLink Properties* window should appear. In the Link <u>T</u>ext box, type `diameter`. Then click on ⬇ near the Book <u>M</u>ark box and select **ANSWER** (or type **ANSWER** in the box). Finally, click on OK. The worksheet should now contain a green <u>diameter</u>. You will need to delete the old "diameter". Try clicking on <u>diameter</u>. The cursor should move to the last equation in the worksheet where we placed the bookmark **ANSWER**.

Try adding a hyperlink to a different worksheet. First create a new worksheet say *shell.mws*, which contains a description of the shell method. Then attach a hyperlink to the phrase "shell method" in the original worksheet.

11. OVERVIEW OF PACKAGES

In Chapters 6 and 7 we needed the *plots* and *linalg* packages. In this chapter we give a brief description of the main functions in some of the other packages. Remember, a package must be loaded with the `with` command. To see a list of the available packages try

```
>  ?index[packages]
```

11.1 Numerical approximation

The numerical approximation package is *numapprox*. Remember to first type

```
>  with(numapprox);
```

Functions include

`chebyshev`	Chebyshev expansion
`hornerform`	convert into Horner form
`infnorm`	L-infinity norm
`minimax`	best minimax rational approx.
`pade`	Pade approximation

11.2 Combinatorial functions

The combinatorial functions are in the *combinat* package. Functions include

`character`	character table of S_n
`choose`	subsets
`graycode`	graycode order
`multinomial`	multinomial coefficient
`partition`	partitions of a given integer
`permute`	permutations
`randperm`	random permutation
`stirling1`	stirling number of the first kind

11.3 Number Theory

The number theory package is *numtheory*. Functions include:

`bernoulli`   Bernoulli numbers and polynomials

divisors	set of divisors
factorset	set of prime diviors
cfrac	continued fraction expansion
cyclotomic	cyclotomic polynomial
jacobi	Jacobi symbol
kronecker	inhom. Diophantine approx.
legendre	Legendre symbol
mcombine	Chinese remainder theorem
minkowski	hom. Diophantine approx.
phi	Euler phi-function
primroot	primitive root
sigma	sum of divisors
sum2sqr	sum of two squares
tau	number of positive divisors

11.4 Orthogonal polynomials

The orthogonal polynomial package is *orthopoly*.

G(n,a,x)	Gegenbauer polynomial
H(n,x)	Hermite polynomial
L(n,x)	Laguerre polynomial
L(n,a,x)	generalized Laguerre polynomial
P(n,x)	Legendre polynomial
P(n,a,b,x)	Jacobi polynomial
T(n,x)	Chebyshev polynomial (first kind)
U(n,x)	Chebyshev polynomial (second kind)

11.5   Statistics

The *stats* package has seven subpackages:

anova	—	analysis of variance
describe	—	data analysis
fit	—	linear regression
random	—	random numbers with a given distribution
statevalf	—	numerical evaluation of distribution function
statplots	—	statistical plotting
transform	—	data manipulation

The following function is available at the top level.

```
importdata(filename,n)
```
Imports data from a file into *n* streams.

Each subpackage must be loaded separately. For instance, to load the *anova* (analysis of variance) subpackage, type

```
>   with(stats[anova]);
```

11.6   Student calculus

The *student* package contains many functions to help the calculus student solve problems step-by-step. In Section 5.7.2 we used the functions **changevar**, **intparts** to do some integration problems. The package also includes the following functions:

completesquare complete the square

distance	distance between two points
Doubleint	double integral
leftbox	plots Riemann sum (also see **middlebox**, **rightbox**)
makeproc	converts expression to function
midpoint	midpoint of two points
showtangent	plots a function together with its tangent at a given point
simpson	Simpson's rule
trapezoid	the trapezoidal rule

11.7   Other packages

DEtools	differential equations tools
Domains	create domains of computation
GF	Galois Fields
GaussInt	Gaussian Integers
LREtools	linear recurrence relations
combstruct	combinatorial structures
difforms	differential forms
finance	financial mathematics
genfunc	rational generating functions
geometry	Euclidean geometry
grobner	Grobner bases
group	finitely-presented groups
inttrans	integral transforms
liesymm	Lie symmetries
logic	Boolean logic
networks	graph networks
padic	p-adic numbers

plottools	basic graphical objects
powseries	formal power series
process	(Unix)-multi-processing
simplex	linear optimization
sumtools	indefinite and definite sums
tensor	tensors in General Relativity
totorder	total orders on names

12. GLOSSARY OF COMMANDS

@   Function composition operator

SYNTAX: **f@g**

Gives the composition of the functions f and g.

EXAMPLE:

```
>  (sin@cos)(x);
```

---

animate   Animation of a 2-dimensional plot
[*plots*]

SYNTAX: **animate(F(x,t),x=a..b,t=c..d)**

Animation of $F(x,t)$ on the interval $[a,b]$ with frames $c \leq t \leq d$.

EXAMPLE:

```
>  with(plots):
   animate(sin(x*t),x=-10..10,t=1..2);
```

---

animate3d   Animation of a 3-dimensional plot
[*plots*]

SYNTAX: **F(x,y,t),x=a..b,y=c..d,t=p..q**

Animation of $F(x,y,t)$ for $a \leq x \leq b$, $c \leq y \leq d$

121

with frames $c \leq t \leq d$.

EXAMPLE:
```
>  with(plots):
   animate3d(cos(x+t*y),x=0..Pi,y=-Pi..Pi,
   t=1..2);
```

---

assign Assignment of solution sets

SYNTAX: `assign(S)`

Assigns the variables given in the set S.

Example:
```
>  S:={y=-1,x=2}:  assign(");  x,y;
```

---

asympt Asymptotic expansion

SYNTAX: `asympt(f(x),x,n)`

Gives the asymptotic expansion to order n of $f(x)$
as $x \to \infty$.

EXAMPLE:
```
>  asympt(GAMMA(x)^2/GAMMA(2*x)*4^x
   /sqrt(Pi),x,3);
```

---

changevar Performs a substitution in an
[*student*] integral

SYNTAX: `changevar(u=g(x),int(f(x),x),u)`

Performs the substitution $u = g(x)$ on the given
integral.

EXAMPLE:
```
>  with(student):  Int(x^2/sqrt(1-x^6),x):
>  changevar(u=x^3,",u);
```

---

coeff Coefficient in a polynomial

SYNTAX: `coeff(p(x),x,k)`
Returns the coefficient of x^k in the polynomial $p(x)$.
EXAMPLE:
```
> expand((1+x+x^2)^10): coeff(",x,10);
```

---

collect   Collect coefficients of like powers

SYNTAX: `collect(expr,x)`
Write the expression as a polynomial in x.
EXAMPLE:
```
> (x+1)^3*y-(y+1)^3*x: collect(",x);
```

---

combine   Combine terms

SYNTAX: `Combine(expr)`
Combines terms in the expression.
EXAMPLE:
```
> combine(sqrt(x+2)*sqrt(x+3));
```

---

contourplot 2-dimensional contour plot

SYNTAX: `contourplot(f(x,y),x=a..b,y=c..d)`
Produces level curves of the function $f(x,y)$ with x, y in the specified ranges.
EXAMPLE:
```
> with(plots): contourplot(sin(x*y),
   x=0..Pi, y=0..Pi);
```

---

convert   Convert data type

SYNTAX: `convert(expr,type)`
Converts the expression to the new *type*.
EXAMPLE:

123

```
> series(sqrt(1-x),x,4):
  convert(",polynom);
```

---

degree   Degree of a polynomial

SYNTAX:`degree(p(x),x)`
Returns the degree of the polynomial in x.
EXAMPLE:
```
> degree((x+y)^6*(y-x^2)^10,x);
```

---

denom   Denominator of an expression

SYNTAX:`denom(expr)`
Returns the denominator of the expression.
EXAMPLE:
```
> denom((x*sin(x)-cos(x))/x^2);
```

---

diff   Differentiation

SYNTAX:`diff(z,x)`
Returns the (partial) derivative $\left(\frac{\partial z}{\partial x}\right) \frac{dz}{dx}$.
EXAMPLE:
```
> diff(sin(x^2*y),x);
```

---

display   Displays a list of plots

SYNTAX:`display(L)`
Displays the plot structures in the list L.
EXAMPLE:
```
> with(plots): P1:=plot(sin(x),x=0..Pi,
  style=POINT): P2:=plot(x,x=0..Pi):
> display([P1,P2]);
```

---

dsolve   Solve ord. differential equations

SYNTAX: `dsolve(deqn,function)`
Solves the given differential equation for the unknown function.
EXAMPLE:
```
> dsolve(diff(y(x),x$2)-y(x)=sin(x),
  y(x));
```

---

evalf Evaluate using floating-point arith.

SYNTAX: `evalf(expr,n)`
Evaluate the **expression** to n digits.
EXAMPLE:
```
> evalf(exp(-Pi),20);
```

---

expand Expand an expression

SYNTAX: `expand(expr)`
Expands the **expression**.
EXAMPLE:
```
> expand((2*x+1)*(3*x-5));
```

---

factor Factor a polynomial

SYNTAX: `factor(p)`
Factors the polynomial p.
EXAMPLE:
```
> factor(x^3+x^2*y-x*y^2-y^3);
```

---

floor Greatest integer function

SYNTAX: `floor(r)`
Returns the greatest integer less than or equal to r.

EXAMPLE:
```
> floor(-11/3);
```

---

fsolve   Solve using floating-point arith.

SYNTAX: `fsolve(eqns,vars)`
Finds an approximate solution to the given set of equations.
EXAMPLE:
```
> fsolve(cos(x)=x/2,x);
```

---

ifactor   Prime factorization of an integer

SYNTAX: `ifactor(n)`
Computes the prime factorization of the integer n.
EXAMPLE:
```
> ifactor(999);
```

---

implicitplot   2-dim. plot of a function defined
[*plots*]   implicitly

SYNTAX: `implicitplot(f(x,y)=c,x=a..b,`
      `y=c..d)`
Plots the set of points (x,y) satisfying $f(x,y) = c$ in the indicated ranges.
EXAMPLE:
```
> with(plots):
  implicitplot((x^2)^(1/3)+(y^2)^(1/3)
  =1, x=-1..1, y=-1..1);
```

---

implicitplot3d   3-dim. plot of a function
          defined implicitly

SYNTAX: `implicitplot3d(f(x,y,z)=c,x=a..b,`
      `y=c..d,z=e..f)`

Plots the set of points (x, y, z) satisfying $f(x, y, z) = c$ in the indicated ranges.
EXAMPLE:
```
> implicitplot3d(x^2+y^2+z^2=1,x=-1..1,
    y=-1..1,z=-1..1);
```

---

int   Compute an integral

SYNTAX: `int(f(x),x)`
Computes $\int f(x)\,dx$.
SYNTAX: `int(f(x),x=a..b)`
Computes the definite integral $\int_a^b f(x)\,dx$.
EXAMPLE:
```
> int(x^2/sqrt(1+x^2),x=1..sqrt(3));
```

---

isolve   Integer solutions to equations

SYNTAX: `isolve(eqns,var)`
Finds integer solutions to the given set of equations (if they exist).
EXAMPLE:
```
> isolve({x^3+x*y=2,x^2+y^2=2},{x,y});
```

---

latex   Convert to LaTeX

SYNTAX: `latex(expr)`
Converts the expression into LaTeX.
EXAMPLE:
```
> latex(Int(1/x,x));
```

---

lhs   Left-hand side of an equation

SYNTAX: `lhs(eqn)`
Gives the left-hand side of the given equation.

EXAMPLE:
```
> e:=x^2+y^2=r^2:  lhs(e);
```

---

limit   Compute a limit

SYNTAX: `limit(f(x),x=a)`
Computes the limit $\lim_{x \to a} f(x)$.
EXAMPLE:
```
> limit((cos(x)-1)/x^2,x=0);
```

---

normal   Normalize a rational function

SYNTAX: `normal(expr)`
Simplifies the **expression** by clearing common factors.
EXAMPLE:
```
> normal((1-q^7)*(1-q^6)/(1-q^2)/(1-q));
```

---

numer   Numerator of an expression

SYNTAX: `numer(expr)`
Returns the numerator of the expression.
EXAMPLE:
```
> numer((x*sin(x)-cos(x))/x^2);
```

---

op   Extracts operands of an expression

SYNTAX: `op(expr)`
Converts the **expression** into a list of operands.
SYNTAX: `op(n,expr)`
Extracts the n-th operand in the **expression**.
EXAMPLE:
```
> w:=x^3+x*y+y:  op(w);  op(2,w);
```

`plot`   2-dimensional plot of a function

SYNTAX: `plot(f(x),x=a..b)`
Plots the function $y = f(x)$, $a \leq x \leq b$.
EXAMPLE:
```
> plot(x*sin(x),x=0..Pi);
```

`plot3d`   3-dimensional plot of a function

SYNTAX: `plot3d(f(x,y),x=a..b,y=c..d)`
Plots the function $z = f(x,y)$, $a \leq x \leq b$, $c \leq y \leq d$.
EXAMPLE:
```
> plot3d(sin(x*y),x=0..Pi,y=0..Pi);
```

`polarplot`   Plots a polar curve
[*plots*]

SYNTAX: `polarplot(f(t),t=a..b)`
Plots the polar curve $r = f(\theta)$, $a \leq \theta \leq b$.
EXAMPLE:
```
> with(plots):
    polarplot(sin(t),t=0..2*Pi);
```

`product`   Find the product

SYNTAX: `product(f(i),i=a..b)`
Computes the product $\displaystyle\prod_{i=a}^{b} f(i)$.
EXAMPLE:
```
> product((a+i-1),i=1..6);
```

`radsimp`   Simplify radicals

SYNTAX: `radsimp(expr)`
Simplify the **expr**ession containing radicals.
EXAMPLE:
```
> radsimp(sqrt(3)*sqrt(15));
```

`rationalize` Rationalize the denominator

SYNTAX: `rationalize(expr)`
Rationalize the denominator in the **expr**ession.
EXAMPLE:
```
> (1+sqrt(2))/(sqrt(2)-sqrt(3)):
  rationalize(");
```

`rhs`   Right-hand side of an equation

SYNTAX: `rhs(eqn)`
Gives the right-hand side of the given equation.
EXAMPLE:
```
> e:=x^2+y^2=r^2:  rhs(e);
```

`seq`   Creates a sequence

SYNTAX: `seq(f(i),i=a..b)`
This creates the sequence $f(a), f(a+1), \ldots, f(b)$.
EXAMPLE:
```
> seq(x+(y-x)*i/4,i=0..4);
```

`simplify`   Simplify an expression

SYNTAX: `simplify(expr)`
Simplifies the **expr**ession.
EXAMPLE:
```
> simplify((sin(x)+cos(x))^2);
```

---

solve   Solve equations

SYNTAX: `solve(eqns,var)`
Finds solutions to the given set of equations (if they exist).
EXAMPLE:
```
> solve({x^2+x*y-y=17,y^2-x-y=9},{x,y});
```

---

spacecurve   Plot spacecurve
[*plots*]

SYNTAX: `spacecurve([f(t),g(t),h(t)],`
           `t=a..b);`
Plots the space-curve parametrized by $x = f(t)$, $y = g(t)$, $z = h(t)$, $a \le t \le b$.
EXAMPLE:
```
> with(plots):
spacecurve([sin(t),cos(t),t,t=0..2*Pi]);
```

---

subs   Substitute into an expression

SYNTAX: `subs(x=a,expr)`
Replaces x by a in the expression.
EXAMPLE:
```
> t^2+t+1: subs(t=1+sqrt(5),");
```

---

sum   Summation

SYNTAX: `sum(f(i),i=a..b)`
Computes the sum $\displaystyle\sum_{i=a}^{b} f(i)$.
EXAMPLE:
```
> sum(i^2,i=1..100);
```

`taylor`	Taylor series

SYNTAX: `taylor(f(x),x=a,n)`
Computes the Taylor series expansion to order n of the function $f(x)$ near $x = a$.
EXAMPLE:
`> taylor(tan(x),x=0,10);`

`value`	Value of an inert expression

SYNTAX: `value(expr)`
Computes the value of the inert `expression`.
EXAMPLE:
`> Int(1/x,x): value(");`

13. FURTHER READING

Below is a list of recent books on MAPLE V.

Introductory books

Heck, A., *Introduction to Maple*, Springer
-Verlag, 1995.

Heal, K.M., Hansen, M.L., and Rickard, K.M., *Maple V Learning Guide*, Springer-Verlag, 1996, 269 pages.

Reference books

Corless, R., *Essential Maple: A Guide for Scientific Programmers*, Springer-Verlag, 1995, 218 pages.

Monagan, M.B., Geddes, K.O., Labahn, G., and Vorkoetter, S., *Maple V Programming Guide*, Springer-Verlag, 1996, 379 pages.

Redfern, D., *The Maple Handbook – Maple V Release 4*, 3rd edition, Springer-Verlag, 1996, 504 pages.

Redfern, D., *The Practical Approach Utilities for Maple – Maple V, Release 3*, Springer-Verlag, 1995, 328 pages.

Maple and Calculus

Bauldry, W.C. and Fiedler, J.R., *Calculus Projects with Maple V*, 2nd edition, Brooks/Cole, 1996.

Cheung, C.K., Murdoch, T., and Keough, G.E., *Exploring Multivariable Calculus with Maple*, Wiley, 1996.

Fattahi, A., *Maple V Calculus Labs*, 2nd edition, Brooks/Cole, 1996.

Hagin, Frank G., and Cohen, Jack K.,*Calculus Explorations with Maple*, Prentice Hall, 1995.

Harris, K., and Lopez, R., *Discovering Calculus with Maple*, 2nd edition, Wiley, 1995.

Maple and Differential Equations

Bugl, P., *Explorations in Differential Equations using Maple*, Prentice Hall, 1995, 149 pages.

Coombes, K.R., Hunt, B.R., Lipsman, R.L., Osborn, J.E., and Stuck, G.J., *Differential Equations with Maple*, Wiley, 1996.

Maple and Linear Algebra

Bauldry, W.C., Evans, B., and Johnson, J., *Linear Algebra with Maple*, Wiley, 1995.

Maple, Science and Engineering

Beltzer, A.I., *Engineering Analysis with Maple /Mathematica*, Academic Press, 1995, 282 pages.

Gander, W. and Hrebicek, J., *Solving Problems in Scientific Computing Using Maple and MATLAB*, Springer-Verlag, 1995, 168 pages.

Greene, R.L., *Classical Mechanics*, Springer-Verlag, 1995, 168 pages.

Horbatsch, M., *Quantum Mechanics Using Maple*, Springer-Verlag, 1995, 331 pages.

Karian, Zaven A. and Tanis, E.A., *Probability and Statistics Explorations with Maple*, Prentice Hall, 1995.

Robertson, J., *Engineering Mathematics with Maple*, McGraw-Hill, 1995.

INDEX

135

137

140